Understanding Electricity

Joyce O. Rasdall
George W. Smith, Jr.

The American Association for Vocational Instructional Materials (AAVIM) is a nonprofit association embracing universities, colleges and divisions of vocational education.

AAVIM's mission is to prepare, publish and distribute quality instructional materials for more effective teaching and learning.

Direction of the organization is provided by representation from member states and agencies. AAVIM also collaborates with teacher organizations, government agencies and industry to provide excellence in instructional materials.

AAVIM Staff

Karen Seabaugh	Director
Loren Dean Roberts	Shipping Supervisor
Brenda Bishop	Customer Service
Kim Butler	Business Manager
Vicki J. Eaton	Art Director/
	Production Coordinator
Scott McKinney	Technical Writer/Editor

Understanding Electricity Editorial Staff

George W. Smith, Jr.	Editor/Art Direction
Joyce O. Rasdall	Revision Author
James A. Anderson	Computer Art
Carol Herring	Computer Layout
Joseph L. Tramnmell, Jr.	Computer Graphics
Bill Staton	Computer Graphics

For information about other AAVIM instructional materials, to place an order or request a free catalog, please write or contact:

AAVIM
220 Smithonia Road
Winterville, Georgia 30683-9527
Phone: (706) 742-5355
Fax: (706) 742-7005
www.aavim.con

ISBN: 0-89606-357-7

Revision Author

Joyce Oliver Rasdall is a professor of Consumer and Family Sciences at Western Kentucky University.

She holds the Ph.D. degree from The Ohio State University where she was an American Home Economics Association Fellow, the M.S. degree from the University of North Carolina-Greensboro and a B.S. degree from Western Kentucky University. A charter member of the National Electric Safety Foundation board and the Blue Grass Chapter of Electrical Women's Round Table, she is a former member services specialist with an electric utility.

Dr. Rasdall periodically reviews electric product safety standards as a member of the Consumer Advisory Council of Underwriters Laboratories. Dr. Rasdall was awarded the Annual Leader of the Year Award by the American Association of Family and Consumer Sciences, the Association of Home Equipment Educators' Annual Outstanding Paper Award (3 times), and a variety of national environmental/energy-responsible design competition honors.

The author wishes to thank Charles O. Shults, engineer, for planting the seeds of electricity, its power and safety for improving quality of life among families, for creating jobs, and for building functional communities.

Appreciation is extended to Martha Jenkins for her variant leadership and service on behalf of the mission of home economics and family and consumer sciences to influence the author and others. The author genuinely appreciates the role of her family in this endeavor as well.

Reviewers and Consultants

A list of all reviewers, consultants, and contributors of information and photographs can be found on page 90.

Foreword

This publication is the result of the combining and updating of information from two AAVIM texts dealing with basic understanding of a power that effects every person in our country every day--electricity. The two AAVIM titles are *Understanding Electricity and Electrical Terms* and *Electric Energy*.

The original manuscript, as well as the revision for the second and third editions of *Understanding Electricity and Electrical Terms,* were written by G.E. Henderson.

Henderson, the first Executive Director of AAVIM and Professor Emeritus, The University of Georgia died in 1997. His desire and talent for explaining difficult concepts with visuals and in terms that are understandable has been an inspiration and challenge to the writers and illustrators of this edition.

Those of us who worked with Mr. Henderson will always be indebted to him for his kindly manner and his insistence on accuracy. This edition is a monument to his continuing legacy to AAVIM and the people who had the pleasure of working with him through his years in education.

The revisions for the fourth and fifth edition were the work of the late J.Howard Turner, former Editor and Coordinator of AAVIM. Dr. Richard M. Hylton, former Director of AAVIM, was the editor of the revisions in the Sixth Edition. George W. Smith, Jr., Director, AAVIM, was the editor for the seventh and eighth Editions.

Electric Energy was written by Dr. W. Harold Parady, former Executive Director, AAVIM and J. Howard Turner, former editor, AAVIM with assistance provided by Ivan L. Winsett, former Executive Director of the Georgia Electrification Council.

Recognition for technical assistance in production of this publication is extended to a number of individuals and companies whose names appear on page 90 of this book. Special recognition is extended to Dr. Joyce O. Rasdall, Professor of Consumer and Family Sciences at Western Kentucky University. Dr. Rasdall is recognized as the revision author of this publication. Without her interest, input and tireless efforts to obtain updated and accurate information, production of this edition would have been impossible.

Contents

Introduction

Photo: James W. Strawser

Figure 1. Electric energy costs are important to all consumers. We should all know more about this essential source of energy we all use every day.

The use of electricity has grown to the extent that an increasing portion of our family expenditures is used to pay for this source of energy. Sometimes you hear a person speak of paying for so many "watts" or "hours" of electricity, when what is really meant is "kilowatt-hours." Some people have trouble understanding electrical terms. Words such as "volts," "watts" and "amperes" are not clear to them.

No doubt such terms have been noticed on the nameplates of electrical equipment. These nameplates give information about the electrical capacity of the equipment and the conditions under which they operate.

Any time a service or commodity affects a consumer's or organization's bank account, the bill payer soon becomes more interested in the amount used (Figure 1).

Electricity: how it works; types of uses, how it is used safely; how it is generated; how it is transformed; transmitted and distributed; how to measure the energy used; and electrical costs are all discussed in this publication. In addition, discussions about electrical safety, emerging technology, effective management of electrical consumption, and the necessity of protecting the environment will be integrated into the material presented on the following pages.

Objectives

From the study of this book, the reader will be able to do the following:
- Describe electric energy and explain how it works.
- Explain how electricity supports quality of life.
- Define common electrical terms, understand and explain their relationships.
- Understand how electricity is used effectively for light, power, heat, electronics, controls, and other applications.
- Understand the importance of electrical safety in the home, the workplace, and the community.
- Determine the amount of electric energy used by selected electric devices during hours of use.
- Compute the cost of electric energy used by electric devices.
- Develop personal energy and environmentally responsible attitudes and behaviors.
- Expand personal awareness of emerging technology as applied to electrical products.

Other educational materials relevant to electricity available from AAVIM:
- *Electric Motors: Selection, Protection, Drives*
- *Electrical Controls*
- *Residential Wiring*
- *How Electric Motors Start and Run*

Supplemental workbooks, instructor's guide and various software products are available for most of these titles. Contact AAVIM for availability and prices of these and other resources.

Figure 2. Electricity drives the technological equipment we find so essential to our lives at work, in our homes, and throughout our communities.

Understanding Electricity & Common Electrical Terms

Electricity is of great importance in our daily activities whether it is used in homes, workplaces, or communities. The purpose of this section is to develop a basic understanding of electricity and electrical terms which supports convenient, safe, effective uses of electrical energy. Upon successful completion of this section, **the reader will be able to describe electricity and identify and define the electrical terms listed in the "Important Terms and Phrases" box** found on this page.

Electricity and Electrical Terms are discussed under the following headings:
   A.   What is Electricity?
   B.   Understanding Common Electrical Terms.

Important Terms and Phrases

Alternating current	Fusetron	Reactive current
Ampere	Fusestat	Resistance
Circuit	GFCI	SHPF
Circuit breaker	Hertz	Semi-conductor
Conductor	Horsepower	Series circuit
Copper	Insulator	Short circuit
Current	Kilowatt-hour	Single-phase
Delivery (hot) conductor	Lightning	Three-phase
Direct current	Nameplate	3-Wire circuit
Electric circuit	Neutral conductor	3-Wire service
Electric current	Ohm	Variable speed drives
Electrical outage	Overloaded circuits	Volt
Electrical pressure	Parallel circuit	Voltmeter
Electricity	Phase converter	Wattage
Electrons	Phase service	Watts
Energy guide	Power factor	Watt meter
Fuse		

A. What is Electricity?

Electricity is the flow of electrons from one atom to another. This flow of electrons is controlled in an **electric circuit**. Electricity is a convenient and controllable form of energy that can easily be converted to other forms such as heat, light, sound, and motion (mechanical motion).

An understanding of the atom is basic to understanding electricity.[1] The center (nucleus) of an atom contains one or more protons. Each proton has a positive (+) charge (Figure 1-A-1). Scientists believe that protons cluster together to form the nucleus of an atom (Figure 1-A-2).

Electrons are arranged in orbital layers around the nucleus (Figure 1-A-3). They move at high speeds. Electrons travel around the nucleus at the speed of light (186,000 miles per second.) Each electron is negatively (-) charged. This causes it to be attracted to the positive (+) proton.

In each neutral atom, the number of electrons and the number of protons are exactly equal. In this manner, the atom is "balanced." For example, a hydrogen atom consists of one proton and one electron (Figure 1-A-4). Another example is **copper**. An atom of

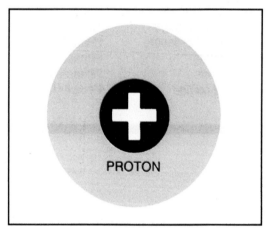

Figure. 1-A-1. The center of an atom contains protons which are positively (+) charged.

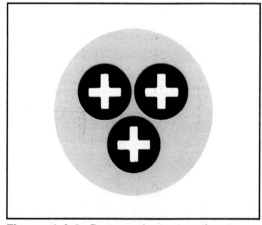

Figure. 1-A-2. Protons cluster together to form the nucleus of an atom.

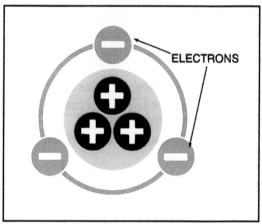

Figure. 1-A-3. One or more electrons travel in orbit around the nucleus of an atom.

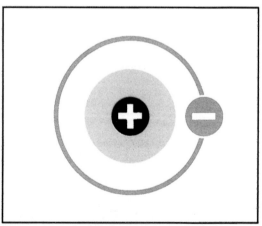

Figure. 1-A-4. A balanced or neutral atom such as the hydrogen atom has the same number of electrons revolving around the nucleus as it does protons in the nucleus.

[1]To present the electron concept of electricity in simple form, selected compromises have been made with the technical facts accepted by physicists. Examples are the omission of neutrons, any indication of the different electron orbits, and the relative size of the nuclei electrons. The atoms illustrated are symbolic rather than indicative of the atomic structure of any particular substance.

copper has 29 protons in the nucleus and 29 electrons in layers orbiting around its nucleus (Figure 1-A-5). In substances such as copper, some of the electrons are loosely held, facilitating an electron exchange between copper atoms.[2] Even though electrons have been exchanged, the atoms are still balanced (Figure 1-A-6). Each atom has the same number of protons and electrons. Because the atom has a loosely held electron in its outer orbit and allows for the exchange of electrons, copper is a good conductor of electricity while hydrogen is a good insulator.[3]

It is possible, when enough force is present, for an atom to lose one or more electrons and become unbalanced (Figure 1-A-7). When one or more electrons are forced on another atom (Figure 1-A-8), the recipient atom has a surplus of electrons and is unbalanced with the positive charges of the protons in the nucleus.

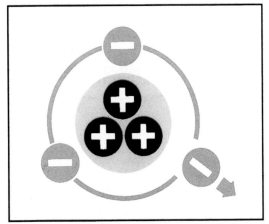

Figure. 1-A-7. It is possible for atoms to lose electrons, thus leaving unbalanced charges.

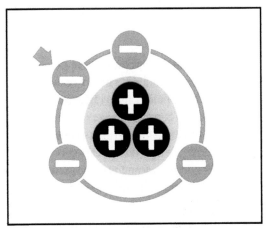

Figure. 1-A-8. Electrons can be forced on atoms making them unbalanced.

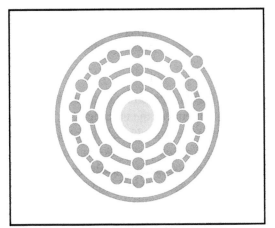

Figure. 1-A-5. Copper has 29 protons in the nucleus and 29 electrons in layers orbiting around its nucleus. The electron in the outer orbit is easily exchanged between copper atoms.[2]

[2]Where the outer orbit has a minority of electrons compared to the outer rings' capacity for electrons, these electrons are loosely held.

[3]Aluminum, similar to copper, has one electron in its outer orbit. Thus it can be an effective electrical conductor in certain well designed circumstances.

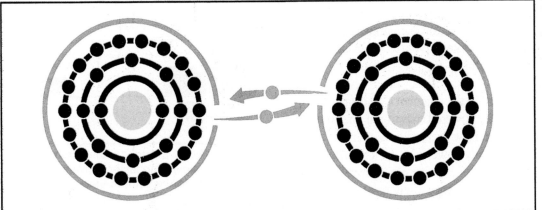

Figure. 1-A-6. Copper atoms may exchange electrons, making copper a good conductor of electricity. The two atoms are still balanced.

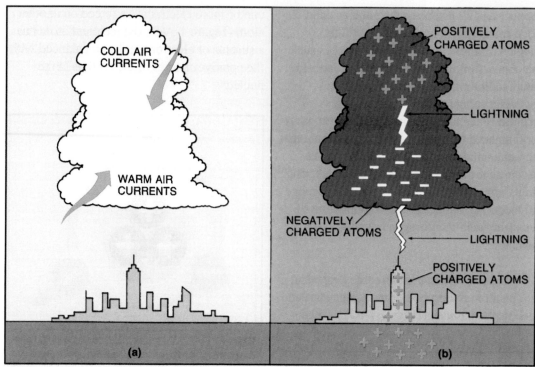

Figure. 1-A-9. Thunderstorms form as air masses move opposite each other in strong currents. Lightning develops as electrons are exchanged between the positively and negatively charged

A similar imbalance of charges is present in clouds during a thunderstorm. A result of this imbalance of electrons is **lightning**. Thunderstorms form as masses of warm air move upward in strong currents. These can be matched by the downward thrust of cold air. As this occurs, some atoms lose electrons and become positively (+) charged. Other atoms gain electrons and become negatively (-) charged. Eventually, an imbalance results and electrons are exchanged between positively and negatively charged atoms in the cloud. This imbalance also sets up an attraction with the positive charges on the ground. This electron exchange between positively and negatively charged atoms causes lightning (Figure 1-A-9).

In contrast to the amount of electrical flow in electrically powered equipment, the amount of electron flow during lightning is tremendously powerful.[4] As a result, each year numbers of people are injured or killed by this natural surge of energy.

When a light is turned on or a vacuum cleaner is operated, the electricity is supplied by the same kind of electron exchange. However, it flows in a controlled manner through a metal **conductor** (wire). This flow of electrons through the conductors is created by the generation of electric power (Figure 1-A-10). Electricity can be generated in a number of different

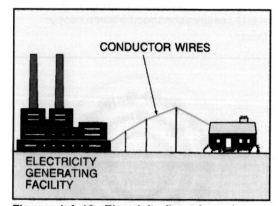

Figure. 1-A-10. Electricity flows from the electricity generating facility through conductor wires to the point of use.

[4]A bolt of lightning can discharge about 100 million volts and heat the air in its path to over 60,000°F at the speed of light (186,000 miles per second).

ways. These methods include internal combustion engines, water power, steam power, solar power, wind power and fuel cells. Approximately 85 percent of all electricity in the U.S. is produced by steam power. Heat for converting water to steam is supplied from fossil fuel (coal, oil and natural gas), nuclear, geothermal (heat from under-ground), and solid wastes. Additional information about the generation of electricity can be found in Chapter 3, "How Electrical Energy is Generated."

B. Understanding Common Electrical Terms

Widespread use of electrical terms affords professionals, tradespeople and consumers a common language with which to discuss electrical energy.

Understanding common electrical terms is discussed under the following headings:
1. How Electrical Energy Moves Through a Circuit.
2. Types of Circuits.
3. Types of Conductors and Insulators and How They Work.
4. How Electric Current is Measured.
5. How Electrical Pressure is Measured.
6. How Resistance to Current Flow is Measured.
7. How Electric Power is Measured.
8. How Electric Energy is Measured.
9. Types of Electric Current.
10. Types of Electric Power.
11. Phase Converters.

1.
HOW ELECTRIC ENERGY MOVES THROUGH A CIRCUIT

A **circuit** is the "path" followed by the electrons (current) from the point where they leave the electricity generating facility until they return to it. This continuous flow of electrons is referred to as **current** or **flow of current**. Generators produce and release current the moment it is needed. Generators deliver the power over a conductor at the speed of light.

There is a continuous flow of electrons from the generator as long as electrical equipment is turned on. Light is the result of current flow. Without current flow, the appliance or motor doesn't run. If an electrical product is turned off or electricity is not produced, the flow of electrons stops. All of the atoms become neutral.

An unscheduled interruption in the availability of energy (or voltage) at its point of use is referred to an an **electrical outage.** The cause may be on the premises as the result of a blown fuse, tripped circuit breaker, wiring problems, or on the utility's power lines (storm damage, wrecks, maintenance, etc.).

Any outage can be inconvenient, disabling, unsafe and costly for consumers and employees alike.

The circuit (Figure 1-B-1) is made up of both a **hot** (delivery) **wire** and a **neutral** (return) **wire**. The reason both wires are needed is to provide one wire over which electrons leave the electricity generating facility and one on which they can flow towards the source.

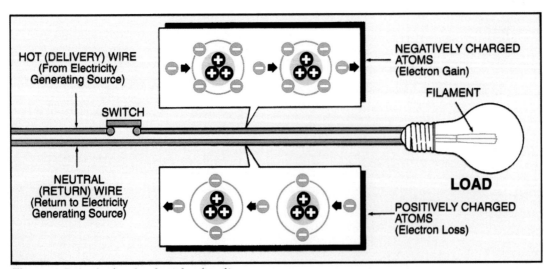

HOT (DELIVERY) WIRE (From Electricity Generating Source)

SWITCH

NEUTRAL (RETURN) WIRE (Return to Electricity Generating Source)

NEGATIVELY CHARGED ATOMS (Electron Gain)

FILAMENT

LOAD

POSITIVELY CHARGED ATOMS (Electron Loss)

Figure 1-B-1. A simple electric circuit.

The neutral (return) wire becomes positively (+) charged as electrons are returned to the electricity generating facility. Electrons from neighboring atoms are attracted and tend to move in and make up the shortage, thus neutralizing the positive atoms. At the same time, the atoms in the hot (delivery) wire are loaded with too many electrons. That portion of the circuit becomes negatively (-) charged. Too many electrons on one side and too few on the other build up **electrical pressure** (voltage). This pressure forces free electrons to move along the hot (delivery) wire and around the circuit to the neutral (return) wire to supply the electron shortage. In turn, the electricity generating facility takes free electrons from the atoms in the neutral (return) wire and forces them on atoms in the hot (delivery) wire. These wires are usually made of copper.

Copper is a good **conductor** of electricity because it gives up electrons freely. Aluminum is used across the country because it is cheaper and lighter. It is not used in house wiring because of expansion, contraction rate, corrosion, and thus potential for unsafe conditions. Copper wires make up the building wiring system.

The circuit includes an electric energy load along with its control device. The light bulb in Figure 1-B-1 represents the load on the circuit. The light bulb filament is a tungsten wire conductor. The same number of electrons that travel through the circuit travel through the filament. The electric energy is converted to light and heat energy. The resistance to the flow of electrons through tungsten causes 80% of the electricity to be converted to heat energy. Only 20% of the electric energy is converted to light energy in a light bulb. Fluorescent tubes operate much cooler because more of the electric energy flowing through them is converted to light than heat.

In making the trip from the electricity generating facility through the lamp and back to the electricity generating facility, no electrons are destroyed. Energy required to push the electrons through the filament is converted to light and heat.

Figure 1-B-2. Fuses and circuit breakers prevent excess current from flowing in the circuit. Fuses are usually cylindrical shaped while circuit breakers are elongated cubes.

The electric current in a circuit passes through either a **circuit breaker** or a **fuse**. Circuit breakers and fuses are protective devices that prevent excess current from flowing in the circuit (Figure 1-B-2). If too many products are plugged into a circuit and/or excess current flows over the circuit, then the breaker flips off or the fuse melts inside from overheating. Thus, circuit breakers and fuses in buildings and equipment prevent wiring from overheating and causing a fire.

An example of unsafe current flow is when the hot wire and neutral wire are shorted (accidentally brought together). A **short circuit** is an unsafe, unwanted path of electricity, usually offering low resistance to the flow of current. Short circuits result from inadequate or improperly installed electric wiring and are a common cause of fire. Although circuit breakers have replaced use of fuses in new houses, fuses continue to protect circuits in many industrial installations.

Two fuse-type protective devices are **fusetrons** and **fusetats**. Both are tamper-resistant, prohibiting dangerous overfusing of circuits. They have a ceramic base mated with a metal screw-in adapter that limits a replacement fuse to the former and correct fuse size in ampere capacity. A **fusetat** permits excessive electrical current over a conductor for a short time, but if the unsafe

Figure 1-B-3. Example of a ground fault circuit interrupter.

condition continues, the wire inside melts to stop the flow of electrons. A **fusestat** can absorb momentary current surges such as those required to start a motor. Motors typically return to running current requirements within the rated limit of the fusestat in a very short time.

Some circuits require additional protection. All outdoor and swimming pool circuits and those used in bathrooms and kitchens for electric appliances, are examples of circuits which require additional protection. These situations pose a greater chance of electrical shock from contact between water, people, and electricity. Protection for these circuits can be provided by a **ground fault circuit interrupter** (GFCI) (Figure 1-B-3). If properly installed, the GFCI is more sensitive than a fuse or circuit breaker.

A tester can be used to test proper installation of a GFCI. A GFCI "opens" the circuit if even a small portion of the electricity is conducted over an "alternative" path (through someone) and is designed to protect people from hazardous shocks. A GFCI is designed to protect humans from electric shock. A GFCI trips when a few milliamperes (5) flow through it and trips in 25/1000 of a second. 50 milliamperes can kill you. Fuses and circuit breakers do not provide this protection.

2.
TYPES OF CIRCUITS

There are two basic types of circuits. They are as follows:

- **Series circuits** (Figure 1-B-4).
- **Parallel circuits** (Figure 1-B-5).

A third type is called series-parallel. It is a combination of the other two.

Series circuits provide for all the current in the circuit to flow through each lamp or appliance on the circuit (Figure 1-B-4). In this type of circuit, if one light bulb (filament) burns out, or the switch is turned off, the circuit is broken. Current flow stops and other light bulbs or appliances that operate on the circuit will not operate.

Parallel circuits provide the same voltage across each load in the circuit and divide current flow through each light bulb or appliance on the circuit (Figure 1-B-5). Current moves through the hot (delivery) wire and divides. A portion of the current passes through the first light bulb and the remaining

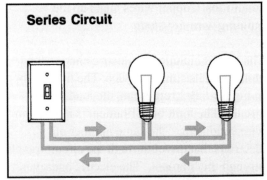

Figure 1-B-4. A series circuit is one in which all of the current flows throughout the entire circuit. Not commonly used today.

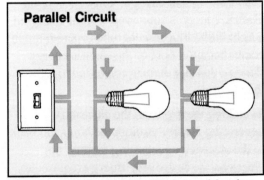

Figure 1-B-5. A parallel circuit provides for dividing the current flow.

portion independently through the second light bulb. Each light bulb is constructed to allow only as much current to pass through it as is needed.

In parallel circuits, if one light bulb burns out or is removed the entire circuit is not interrupted. Current continues to flow through the other light bulbs in the circuit. Most circuits in the home are of this type (fuses, switches and controls are normally used in series in wiring circuits and lights and appliances are in parallel).

There are six conditions inside a building that will break the entire circuit and stop current flow in parallel circuits. They are as follows:

- Main switch off for entire building.
- Circuit switched off manually.
- Circuit breaker or ground fault circuit interrupter automatically tripped or fuse burned out.
- All lamp filaments burned out.
- Switch in "off" position.
- Lamp or appliance unplugged.

3.
TYPES OF CONDUCTORS AND INSULATORS AND HOW THEY WORK

As mentioned earlier, each atom has as many electrons around it as it has protons. Some substances have free electrons, electrons that are held rather loosely. As a result, atoms in these substances can be forced to give up electrons and accept others with relative ease. Since some materials permit electrons to move readily, they are called **conductors** (Table I). Silver, copper, aluminum, most other metals and impure water (rain water, bath water, dish water, swimming pool water, ground water, and body liquids) are examples of good conductors (Figure 1-B-6).

There are other materials that will not conduct electricity, such as glass, cloth, rubber, plastic, porcelain, paper, dry wood and pure distilled water, whose protons hold tightly to their electrons (Figure 1-B-7). They will not

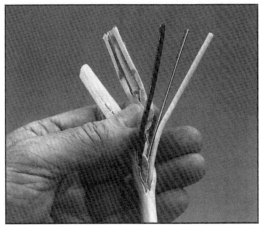

Figure 1-B-7. Insulation covering electric conductors is made of a material which does not exchange electrons. Insulation will not release its own electrons nor receive other free electrons.

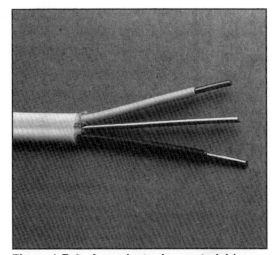

Figure 1-B-6. A conductor is a material (usually wire) that allows electron exchange. A good conductor, such as copper or aluminum, gives up electrons easily. However, copper is better because the electrons in the outer orbit are more loosely held.

TABLE I.		
Examples of Conducting, Semi-Conducting and Insulating Materials		
Conductors	**Semi-Conductors**	**Insulators**
Metals	Nichrome	Distilled water
- Gold	(Nickel and	Dry nonconductors**
- Silver	Chromium	- Cloth
- Copper	Alloy)	- Wood
- Aluminum	Tungsten	- Paper
- Iron	Electro-magnetic coils	- Plastics
- Others		- Rubber
Human body		- Glass
Animals		- Ceramics
Impure water		- Concrete
- Rain water		- Brick
- Groundwater		- Stone
- Household water		Air
- Body liquids*		Fiberglass
- Ponds, lakes, & streams		Porcelain
Earth		
Most wet surfaces		

*The human body and animals contain ions of aluminum, iron, copper, and other conductors.
**If any metallic fibers or components are fabricated in these products, they can become conductors.

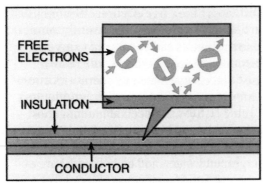

Figure 1-B-8 Insulating materials hold their electrons tightly. They will not let free electrons flow through them.

release their own electrons nor accept other free electrons. They are known as **insulators** or **non-conductors**. If rain or household water spread across surfaces of or penetrate wood, cloth, paper, or other insulators, then electrons can flow through or on damp surfaces of materials (Table I).

Materials which are neither insulators nor conductors are called **semi-conductors**. These substances have resistance to electrical flow because their electrons do not move freely. This resistance results in heat. Nichrome, used in ranges, toasters and clothes dryers, is a semi-conductor. Tungsten filaments produce incandescence (light) and heat because of electrical resistance.

Insulators play an important part in controlling electricity safely and conveniently. Wrapping insulating material around a conductor (wire) prevents it from touching an-

other conductor. This safely keeps current in the proper path. Thus, current is not free to flow to other conductors such as the human body (Figure 1-B-8).

Insulation-covered wires (conductors) which connect lamps and appliances protect users from getting shocked (Figure 1-B-9). They prevent the body from completing an unsafe circuit between bare wires and the ground or other conductor. The insulating material also keeps the hot (delivery) and neutral (return) wires apart, thus, preventing a short circuit.

4.
HOW ELECTRIC CURRENT IS MEASURED

Electrons travel along a conductor at the speed of light (186,000 mps). This is similar to water flowing in a stream or pipe and is known as "**current**." When 6,280,000,000,000,000,000 (6.28 quintillion) electrons pass any particular point in a circuit in one second, this is called one **ampere** (abbreviated as "I" for "Intensity of current"). It may also be referred to as **amperage** (or **amp**). The ampere is the unit of measure of electric current. A 100-watt light bulb is designed to require less than one ampere to work properly.

An instrument called an ammeter measures this current flow (amperes) (Figure 1-B-10).

Photo: James W. Strawser

Figure 1-B-9. Insulation keeps users from being shocked when handling appliance cords.

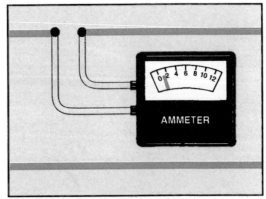

Figure 1-B-10. An ammeter measures the current (electron) flow in a wire. If the ammeter were installed on the lower wire, it would show the same number of amperes. The same amount of current is flowing in each wire.

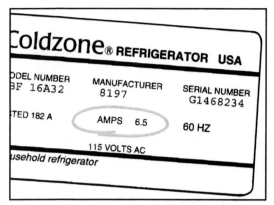

Figure 1-B-11. Capacity of electrical equipment is often given in amperes on the nameplate. This information enables an electrician to connect the correct wire size to equipment and the circuit breaker panelboard.

Current must actually pass through the meter to measure the current. If no current is flowing, the ammeter reads "0."

Since the ampere is a measure of current flow, it is important to be able to understand and describe the current carrying capacity of conductors and other electrical equipment. Ampere (or amperage) rating of most heating and power equipment indicates the amount of current needed to operate the equipment properly. It is generally shown on the **nameplate** (Figure 1-B-11). Nameplates on electrical equipment show their ratings in amperes. This information can be used to determine what size wire to use when installing a special circuit, as for a clothes dryer or an air conditioner. Nameplates are located on appliance frames, behind doors (on appliances), on the side and bottom, or etched into the metal or plastic surfaces of the item.

Figure 1-B-12. Fuses and circuit breakers are rated in amperes to show the maximum electrical load they will carry before they melt or trip.

Different parts of a building's wiring system, such as wall switches, circuit breakers and fuses (Figure 1-B-12), are also rated in amperes. These ratings indicate the maximum amount of current they can safely handle. If more than that amount of current passes through them, they will be damaged from excessive heat. Too much current makes a circuit breaker "trip" or fuse melt inside. Circuit breakers and fuses of the proper size will protect the wiring from overheating (and possible fires) and warn users of overloads.

5.
HOW ELECTRICAL PRESSURE IS MEASURED

As explained earlier, electrons can be forced to move through a conductor in an electric circuit. The **volt** (abbreviated as "E" for "electromotive force") is the unit of measure of electrical pressure, the pressure being applied to force electrons through a circuit. This pressure causes electrons to flow through an appliance or light when connected. This pressure, generally referred to as **voltage**, is available in energized circuits all the time whether electrical equipment is being or used or not.

When an appliance that requires a lot of current is started, voltage in conductors can drop slightly. Consequently, lighting output can dim, television pictures "shrink", sound equipment sound "scratchy" for a moment, and be troublesome to computers and controls. But if wiring is properly sized, the drop will not be enough to affect other equipment. This voltage drop can be detected with a **voltmeter** (Figure 1-B-13).

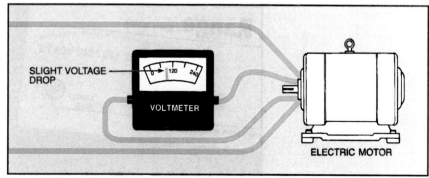

Figure 1-B-13. Voltage should change very little when starting equipment, except in some instances with large users of electricity.

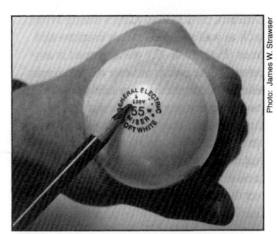

Figure. 1-B-14. Voltage marking on a light bulb.

Circuits used for lighting and for small appliances around a home, farm or business are typically supplied with 115 to 120 volts. The American National Standards Institute (ANSI) specifies 114 to 127 volts. Actual voltage could be a few volts higher or lower. This voltage will serve any equipment that has approximately the same voltage marking stamped on its nameplate. For example, the light bulb in Figure 1-B-14 is stamped "120 volts."

Larger equipment, such as major heating appliances and 1/2-horsepower motors or larger, is designed for use with 240 volts. In fact, many electric ranges use both 120 and 240 volts to help supply the different heat settings needed for surface cooking. Note that the electric range nameplate shown in Figure 1-B-15 indicates "volts 120-240." The lights on control panels and in the oven, clock, and timer on gas and electric ranges use only 120 volts.

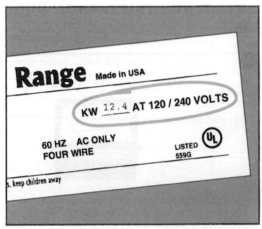

Figure. 1-B-15. Voltage marking on a typical range nameplate.

Some motors are designed so they can be connected for use with 120 volts, or they can be reconnected for use with 240 volts, but not both at the same time. Figure 1-B-16 shows how dual-voltage motors are stamped. With this understanding, the meanings of the electrical terms on the nameplates of electrical equipment become important.

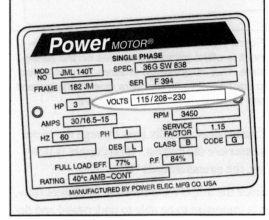

Figure. 1-B-16. Dual-voltage marking on a motor nameplate.

Equipment that operates on one voltage will show a high and low voltage on the nameplate, such as 110-120. This means that any voltage between these two figures should be satisfactory for operating that piece of equipment. The National Electric Manufacturers Association complies with ANSI specifications. Electric utilities utilize ANSI specifications as well.

When equipment does not work satisfactorily, a voltage check by the local power supplier or electrician may show that the circuit voltage is below the voltage shown on the nameplate. If so, heating equipment heats slower, light sources produce less light, and motors can fail to start readily and will develop full power more slowly, if at all. On the other hand, higher than recommended voltages cause lights to burn brighter (and hotter which reduces lamp life), heating equipment to heat more quickly, and motors start quicker and tend to overheat.

On electric range nameplates (Figure 1-B-15) the term 3W or 4W will be seen. This also relates to voltage. 1996 NEC requires

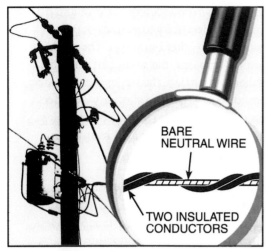

Figure. 1-B-17. When there are three wires, usually both 120 and 240-volt services are available.

four wires (4W). Both supply 120 and 240 volts simultaneously and require 3 wire service.

Consumers can usually identify a 3-wire service to their house by checking the number of wires (conductors) that come from the power pole to the electric meter, or from the meter pole to the house (Figure 1-B-17). This is the standard power supply to houses. If there are three wires (conductors), both 120 and 240 volts are available for use.[5] Count the wires entering the weatherhead

Figure. 1-B-18. Sometimes the wires are twisted together but they can still be counted where they enter the weatherhead. The three wires will not be visible when the house is supplied through underground service.

(Figure 1-B-18). Some users confuse **3-wire service** with **3-phase service**. Three wires running from the power pole to the meter does not mean a customer has 3-phase service. This is usually 3-wire single phase service–two insulated hot (delivery) wires and a bare neutral wire. Three-phase service generally has four wires from the power pole to the meter–three insulated hot (delivery) wires and a bare neutral wire. See Section 10 in this Chapter for further discussion of this topic.

6.
HOW RESISTANCE TO CURRENT FLOW IS MEASURED

Even when the electric current flows through good conductors, some energy is required to force electrons from one atom to the next. The energy required to overcome the resistance is converted into heat. This condition is called **resistance**. An **ohm** (abbreviated as "R" for resistance") is a unit of electrical resistance.

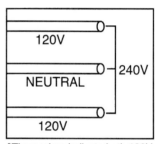

[5] Three wires indicate both 120V and 240V service are available.

There are three factors which determine the amount of resistance (ohms) in a conductor (Figure 1-B-19):

- **The material of which the conductor is made** (Table I). Although gold and silver are efficient conductors, cost and other

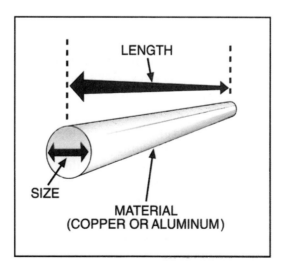

Figure. 1-B-19. The amount of resistance in a conductor is determined by three factors: material, size and length.

TABLE II. **Properties of Effective and Poor Conductors**		
Property	Effective Conductivity	Poor or Unsafe Conductivity
Conductivity of substance	Excellent: Gold, silver, copper Good: Aluminum Fair: Iron, others	Typically other metals, and water, wet surfaces and damp materials.
Diameter of conductor	Sufficiently large to reduce resistance and enhance conductivity	Too small diameter of conductor increases resistance, heat, and dangerous conditions in conductors.
Length of conductor	Short paths of current flow reduce resistance and overheating	Long conductor wires increase resistance, heat, and dangerous conditions in conductor.
Cost/foot	Very High: Gold, silver Moderate: Copper, aluminum Low: Iron, others	

limitations associated with these metals are prohibitive (Table II). Copper and aluminum are commonly used conductors although aluminum can incur more resistance than copper. Other metals are conductors but are less effective than aluminum.

- **The size (diameter or gauge) of the conductor.** The larger of two conductors of the same type and length will offer less resistance to the flow of electrons.
- **The length of the conductor.** It is logical that electrons traveling a shorter distance would meet less total resistance. Resistance in conductors always generates some heat when current flows through the conductor. Table II provides additional information about conductivity and resistance.

The efficient and effective flow of electric current depends on choosing properly sized materials with good conducting qualities whose path is as short as possible. Using material that is excessively long, undersized, with poor conducting qualities, is less effective, more expensive and may create unsafe conditions.

Resistance in conductors is a problem in supplying electricity to various outlets in homes and in other buildings. The more current (amperes) the wires carry and the farther it is carried, the larger the wire size should be. Larger and shorter wire offers less resistance. Less resistance means less loss of energy in the form of heat when current flows in the wire.

Resistance to electrical flow causes some electric energy to be converted to heat energy. The greater the resistance, the less efficient the electrical conduction process which can cause dangerous fires and result in loss of property, personal injury, and even loss of life.

Some substances are neither good conductors nor good insulators. They are called **semi-conductors** and they hold their own free electrons rather tightly, posing resistance to current flow. This results in much heat being given off. An example of such a material is nichrome wire used in heating elements in electric ranges, toasters, clothes dryers, and room heaters (Figure 1-B-20). Of course, heat used in this manner is for a safe, useful purpose. Heat given off by conductors in the wiring system, however, is a complete loss of electricity paid for by the consumer. Excess heat in wiring of buildings is a potential life threatening danger to occupants.

Photo: David Gentry, Warren RECC

Figure 1-B-20. Semiconductors have resistance to electron flow. Nichrome, an alloy of nickel and chromium, is a semiconductor useful in appliances that produce heat.

Resistance in electrical wires and their connected equipment can be measured with an ohmmeter. An easy way to remember the relationship between amp, volts, and ohms is **Ohm's Law** (E=I x R). This law is one of the basic formulas used to quantify (measure) electricity. Expressed as an equation:

$$Volts = Amps \times Ohms$$

As with all equations, this equation can be modified algebraically to read:

$$Amps = \frac{Volts}{Ohms}$$

$$Ohms = \frac{Volts}{Amps}$$

Example:

$$Ohms = \frac{120V}{12 \text{ amps}} = 10 \text{ ohms}$$

(12 amps represents the maximum recommended loading for a 15 amp circuit.)

7.
HOW ELECTRIC POWER IS MEASURED

Electric power is measured in **watts**. It may also be referred to as "**wattage**." Remember that amperes is current flow and volts is electrical pressure. But they are a team when power is discussed. Neither term by itself gives a measure of power for turning motors, or for producing heat or light. For example, if 15,000 volts were available but no free electrons flow, there would be no power. Or, if there are enough free electrons in a circuit to provide a flow of 3,000 amperes, there will be no power unless there is voltage (pressure) to make them flow. But, if 15,000 volts and 3,000 amperes are combined, enough power for a small city could be transmitted.

A unit of measurement that indicates what voltage and amperage are in terms of power can be calculated by multiplying the two together and then multiplying by a unitless fraction called the **power factor**, to produce

a new term, **watt**. Power factor is dependent upon the type of current and the type of device the current is flowing through. See Section 9 for a discussion on the types of current. The power factor is always 1.0 in **direct current** (DC) circuits. In **alternating current** (AC) circuits, the power factor ranges from 0 to 1.0. For an AC electric heater, the power factor will be nearly 1.0. For a small AC motor, the power factor may, at times, be as low as 0.25 depending upon loading.

Thus, for a DC circuit,
Volts x Amperes = **Watts**

For AC circuits when the power factor is 1,
Volts x Amperes = **Watts**

For AC circuits when power factor is calculated,
Volts x Amperes x Power Factor = **Watts**

To show how this works in DC circuits, or AC circuits where the power factor is 1.0, suppose there is a 120 volt circuit in which 1 ampere is flowing. Multiply v x a = w.
120 volts x 1 ampere = **120 watts.**

Now suppose there is another circuit with 240 volts and a current flow of 1 ampere. Multiply v x a = w.
240 volts x 1 ampere = **240 watts.**

In other words, by doubling voltage the watts available (power use) has been doubled with the same amperage flowing in the circuit. This assumes that the resistance is doubled.

For an AC circuit that has a power factor of 0.50 the power would be:

Volts x Amperes x Power Factor = **Watts**
120 volts x 1 ampere x 0.50 = **60 watts**

If either the voltage or amperes of current is doubled, the power is also doubled in AC circuits.

Another component needed to operate motors, fluorescent ballast, etc., is **reactive**

Relationship of Terms

120V ▲ ▼ 12A	**10 ohms** (resistance)
Watts = V x A 1440 = 120 x 12	
Watts = A²R 1440 = (12)² x 10	
Watts = E² ÷ R 1440 = (120)² ÷ 10	

Figure 1-B-21. One horsepower can be figured as roughly equal to 1,000 watts. (Nameplate shown above figures, 4.2 amperes x 240 volts = 1,008 watts.)

current which, combined with the real current determines power factor and results in a total of more than just the volts x amps. Low power factor can overload circuits, conductors, transformers and other equipment.

Voltage is important with equipment that is considered a heavy load. Equipment requiring 240 volts requires a wire only half the size (diameter) of equipment using 120 volts in delivering the same amount of power. Also, with 240V the amps will be $\frac{1}{2}$ that of the 120V circuit. The flow of amps (electrons) in the circuit determines the wire size needed.

Horsepower. Many items of electrical equipment are rated in watts. One exception is electric motors. Most motors are rated in horsepower (Figure 1-B-21). Horsepower can be changed to watts rather easily by figuring approximately 1,000 watts for each horsepower of motor rating if the motor is $\frac{1}{2}$ horsepower (hp) or larger. For motors of less than $\frac{1}{2}$ hp use 1,200 watts per hp. A 5-hp motor will be about 5,000 watts at full load. A $\frac{1}{4}$ hp motor will be about 300 watts if it is pulling a full load.

A theoretical horsepower is 746 watts, but this does not include loss of energy in the motor through electrical losses and friction.

If motors are not fully loaded, they do not develop full rated horsepower, therefore wattage is less. If they are overloaded for a short period of time, they will develop more than their rated horsepower. Then their wattage is more.

Overloaded Circuits. An example of how watts, volts, and amperes relate to each other in a practical situation might be shown in the following example.

A microwave oven is being used on a circuit protected by a **20 amp circuit breaker.** The nameplate on the microwave oven shows 1500 watts and 120 volts (Figure 1-B-22). A coffee maker (nameplate shows 1000 watts and 120 volts) is added to the same circuit (Figure 1-B-22). The circuit breaker is tripped. What caused the circuit breaker to trip off?

To determine why the circuit breaker tripped, first calculate if the combined amperage of the two appliances exceeds the 20 amp breaker protecting the circuit.

Microwave Oven (Figure 1-B-22)
    1500 watts ÷ 120 volts = 12.5 amperes

Coffee Maker (Figure 1-B-22)
   <u>1100 watts ÷ 120 volts =   9.2 amperes</u>
   2600 watts Total   **21.7 amperes**

The total number of amperes (21.7) required by both appliances exceeds the 20 amp circuit breaker protection. This results in an overload on the circuit causing the breaker to trip.

Another example is the use of an iron and hair dryer at the same time.

Hair Dryer
    1200 watts ÷ 120 volts = 10 amperes

Iron
   <u>1300 watts ÷ 120 volts = 10.8 amperes</u>
   2500 watts Total   **20.8 amperes**

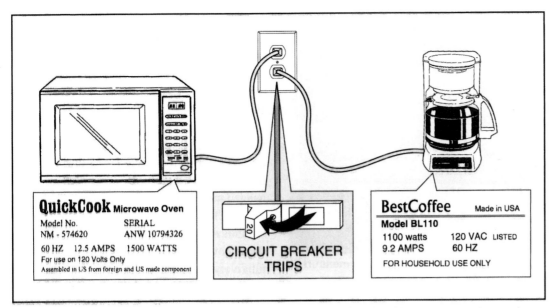

Figure. 1-B-22. The overload on a circuit can be determined by checking the nameplate rating on each piece of equipment using the circuit.

The total amperes (20.8) exceeds the 20 amperes permitted by the 20 ampere circuit breaker causing the breaker to trip off.

Equipment of more than 1,000 watts is often rated in **kilowatts.** A kilowatt is 1,000 watts. The microwave oven in Figure 1-B-22- could have been rated KW 1.50, volts 120 which would be the same as the rating of watts 1,500, volts 120. A micro-wave oven with that much wattage should be on its own special circuit to prevent nuisance fuse blowing or the circuit breaker tripping to off. Watts can be measured with an instrument called a **wattmeter.**

Present day general purpose and small appliance circuits are of 12 gauge wire which should be protected with a 20 ampere circuit breaker or fuse. However, some older homes and buildings may still be wired with the next smaller size wire (14 gauge) for general purpose circuits. Table III provides information about requirements for typical wiring used in residential uses. With 14 gauge wire, 15 ampere fuses or circuit breakers are used.*

*Enclosed fuses and circuit breakers are rated at only 80% of their marked value. Continuous load should be .8 x vxa.

TABLE III. American Wiring Size Requirements for Typical Residential Applications				
Uses	Copper Wire Size	Maximum Ampere Rating	Voltage	Maximum Wattage*
Door Bell, lamp cord	18	10	120	1200
Lighting circuits (no longer permitted in most new house wiring)	14	15	120	1800
Lighting circuits, general purpose circuits, individual equipment circuits such as dishwasher, refrigerator, freezer, automatic washer, small 120 v air conditioner and 120 v wall heater	12	20	120	2400
Water heater	10	30	240	4,500
Clothes dryer, built-in range, built-in oven, small a/c	10	30	120/240	5,000
Electric range, central a/c, electric furnace or heat pump	6	55-60	120/240	13,200 14,400
100 amp service entrance (copper clad or aluminum conductor)	2	100	120/240	22,800
150 amp service entrance (copper conductor)	1	150	120/240	30,000
200 amp service entrance (copper clad or aluminum conductor)	000	200	120/240	48,000

*Recommended maximum wattage should be designed to be 80% of the wattage listed.

Figure 1-B-23. Electrical energy is measured and purchased by kilowatt-hours.

8.
HOW ELECTRIC ENERGY IS MEASURED

What does it cost to operate? This is a question frequently asked, especially when new equipment is being purchased. It is an important question, one that can usually be answered if wattage and electrical energy measurements are understood.

It is customary to buy goods by measured volume (gallons or liters), or by weight (ounces, grams, pounds or kilograms). For example, gasoline or milk is purchased by the gallon, and animal feed or meat by the ounce or pound. Electrical energy is purchased by kilowatt-hours (1,000 watt-hours) (Figure 1-B-23).

In the section "How Electric Power is Measured," a watt-hour is described as the rate of using electric energy or the rate of doing electrical work. For example, when one ampere is pushed through a circuit by 120 volts for one hour, the power is 120 watt-hours. This is approximately the power required to turn a small food mixer for an hour (Figure 1-B-24).

But watts and horsepower do not give a measure of the amount of electric energy a piece of equipment will use. Another unit is needed which includes the time the equipment is "on." This unit is the watt hour (**whr**). It combines the unit of power (**watt**) with time (**hour**) over which electricity is used. It is a relatively small unit. A block of 1,000 is more commonly used. "Kilo" is the metric unit meaning "1,000." One thousand watt-hours equals one **kilowatt hour** (kwh). This is the unit of electric energy used and paid for each month.

The term "kilowatthour" is abbreviated in several ways (Kwh, kwh, KWH, etc.) by manufacturers and utilities. To lessen confusion in this publication, **kwh** will be used in the text when making reference to kilowatthours.

For example, a man is hired to mow the lawn. He might be rated as "1 manpower"; but he is paid only for the hours that he works. He is paid by the hour, or manpower-hour.

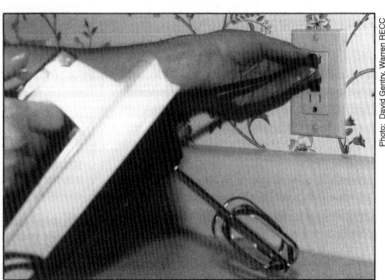

Figure 1-B-24. The power required for an hour's use of a small food mixer is about 120 watt-hours. Watts measure power for heat and light as well as for motors.

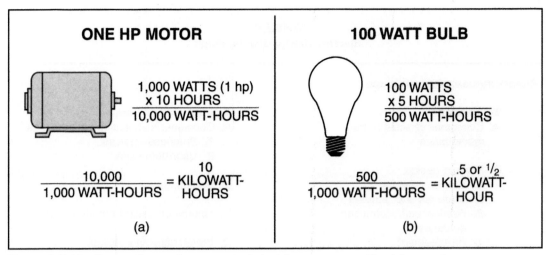

ONE HP MOTOR

$$\frac{1,000 \text{ WATTS (1 hp)}}{\times 10 \text{ HOURS}}$$
$$10,000 \text{ WATT-HOURS}$$

$$\frac{10,000}{1,000 \text{ WATT-HOURS}} = \frac{10}{\text{KILOWATT-HOURS}}$$

(a)

100 WATT BULB

$$\frac{100 \text{ WATTS}}{\times 5 \text{ HOURS}}$$
$$500 \text{ WATT-HOURS}$$

$$\frac{500}{1,000 \text{ WATT-HOURS}} = \frac{.5 \text{ or } 1/2}{\text{KILOWATT-HOUR}}$$

(b)

Figure 1-B-25. How to figure the amount of electric energy used by (a) a one-horsepower motor, and (b) a 100 watt light bulb.

Electric energy is paid for in the same way. Assume operating a 1 horsepower motor fully loaded for 10 hours. Multiplying 1,000 watts (approximately equal to 1 horsepower) by 10 hours, gives an answer of 10,000 watt-hours or 10 kwh. Watt hours are such small units of electrical energy that power suppliers sell them by the thousand. To calculate the kilowatt-hours of electrical energy, divide the number of watt hours by 1,000 (Figure 1-B-25a). Or, simply move the decimal after 10,000.0 watts three places to the left which yields 10.000 or 10 kwh.

A 100 watt light bulb used for five hours is calculated the same way (Figure 1-B-25b). Power for a food mixer is computed in the same way, but note that a mixer is used only a few minutes at a time. Suppose it is used for only 15 minutes (1/4 hour). The energy used may be calculated as shown in Figure 1-B-26.

120 WATT MIXER USE FOR 1/4 HR.

$$15 \text{ MINS.} = 1/4 \text{ HR.}$$
$$120 \text{ WATTS} \times 1/4 \text{ HR.} = 30 \text{ WATT HRS.}$$

$$\frac{30 \text{ WATT-HRS.}}{1000} = .030 \text{ kwh}$$

Figure 1-B-26. Energy used by an electric mixer.

Appliances using electricity are rated through a U.S. government based test for their energy efficiency. Every new refrigerator, electric range, washer, dryer, and water heater must display an **Energyguide** sticker. This sticker will indicate the estimated cost of operation of a particular appliance relative to other similar equipment (Figure 1-B-27). The lower the cost per hour of use, the more energy efficient the appliance.

Cost of operation of electric products depends on the following factors; fabrication

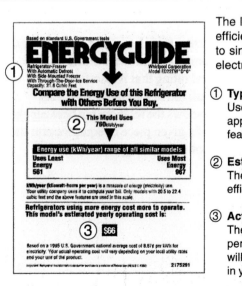

The lower the kwh used, the more efficient the appliance compared to similar items and the less electricity it will take to operate it.

① **Type of Appliance and Capacity.**
Use this information to compare appliances of similar size and features.

② **Estimated Energy Use.**
The lower the number, the more efficient the appliance.

③ **Actual Cost per Year.**
The cost of $66 is based on 8.67¢ per kwh for this example. Your cost will vary depending on the kwh cost in your area.

Figure 1-B-27. The Energyguide sticker provides information about the energy efficiency of a product relative to other similar appliances.

Electricity usage depends on:

1. **Equipment**
   A. Consumer choices in the marketplace

   B. Product design
      1. Size or capacity
      2. Materials and features
      3. Fabrication/construction processes
      4. Arrangement

2. **Building**
   A. Consumer choices in the marketplace

   B. Design (building and space)
      1. Size, volume or square footage
      2. Materials and features
      3. Fabrication/construction processes
      4. Arrangement of space
      5. Orientation to sun and wind

3. **Amount of "on" time**
   A. Consumer's choice/behavioral uses
      1. Purchase choices
      2. User behaviors

   B. Switching type

   C. Volume of space to heat or cool

   D. Product/building design

   E. Weather/room conditions

4. **Utility rates**

materials, design of product, and user behavior which determines "on time". Table IV provides information on electric energy use variables. Cost of equipment operation depends on consumer behavior, energy efficiency, the local electric utility rate and the time used.

The efficiency of electrical heat pump and air conditioning equipment is indicated by a numerical rating that is also found on a sticker affixed to the equipment. This number shows the relationship between the output of a given piece of equipment to the amount of input energy.

The number shown on the sticker is the **SEER** (seasonal energy efficiency ratio). This number is determined by dividing the total estimated cooling in BTU's by the estimated energy consumed in watt hours. The higher the number, the more energy efficient the equipment.

$$\textbf{SEER} = \frac{\text{Estimated Cooling Delivered (BTU's)}}{\text{Estimated Energy Consumed (watt Hrs)}}$$

Since January 1992, this rating number has been required to be 10 or above on all air conditioning equipment.

There is also an efficiency factor for the heating season for electric heat pumps known as **SHPF** (season heating performance factor). This rating number should be at least 7.

9.
TYPES OF ELECTRIC CURRENT

Current has been previously discussed as flowing steadily in a circuit. This is called **direct current**. On most nameplates, it is abbreviated as "DC." DC is the type of current used in flashlights, portable radios, cameras, boats, and automobiles (Figure 1-B-28).

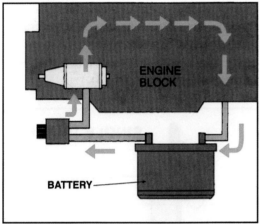

Figure 1-B-28. In an automobile engine direct current flows in one direction only.

When electricity was first made available, it was all direct current. DC was found to have a serious disadvantage; it was difficult to deliver over long distances. Direct current is used for home and industrial applications today by converting alternating current to direct current. This conversion takes place inside of appliances such as televisions, radios, stereos and computers.

The current which flows through lights, refrigerators and other equipment in homes, flows with fluctuating voltage (Figure 1-B-29b). It is called **alternating current** (AC) because voltage fluctuates from 0 to 120 volts positive, then to 0 to 120 volts negative.

This current flow with voltage fluctuations is called a **cycle**. In the U.S., 60 of these cycles occur each second, thus, the term 60-cycle AC. The term for "cycles per second" is **hertz** or "**HZ**" (Figure 1-B-30).

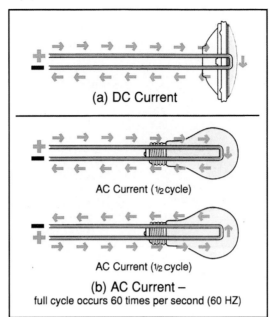

Figure 1-B-29. (a) DC current always flows in one direction, unchanging, as through this automotive headlight bulb. (b) AC current flows in one direction for one-half cycle, then reverses direction and flows in the other direction for one-half cycle, completing a cycle. This action occurs 60 times per second (60HZ).

All power suppliers in the U.S. control the power to exactly 60 hertz. Because of this control, electric clocks keep accurate time. If there should be more cycles per second, clocks would gain time; if fewer, clocks would lose time.

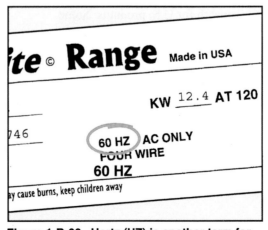

Figure 1-B-30. Hertz (HZ) is another term for cycles per second.

Figure 1-B-31. This drill is powered by a battery pack that is recharged by alternating current.

Most equipment for homes is designed for alternating current only. But there are a few items of equipment that will operate on either AC or DC. Examples of these are standard light bulbs and some portable electric drills. Battery powered hand tools are widely used today. The battery packs are recharged by alternating current (Figure 1-B-31).

Before connecting an ammeter, voltmeter or wattmeter to an alternating current circuit, always check the nameplate to see that the meter is designed for use with alternating current.

Because life is increasingly global today, travelers to some countries can be inconvenienced when their 50 cycle current is a mismatch for U.S. products. Some countries do not have 120 volt electric service. No American 120 volt appliance should ever be used on 240 volt electric service overseas. American travelers can secure appropriate adapters and transformers to facilitate American appliances being operable overseas.

10.
TYPES OF ELECTRIC POWER

The terms **single-phase** and **three-phase** are confusing to most consumers. Some users see three wires running from the power pole to their meter (Figure 1-B-32) and think this means they have three-phase service. This is usually 3-wire single-phase service. This service provides both 120 and 240 volts as discussed under the heading, "How Electrical Pressure is Measured" (Section 5). It has two hot (delivery) wires and a neutral (return) wire. Three-phase service generally has four wires from the power pole to the meter, three (delivery) wires and a neutral (return) wire (Figure 1-B-33).

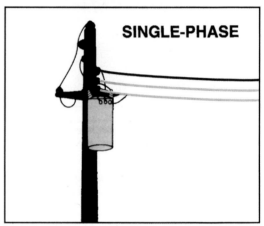

Figure 1-B-32. Single-phase power line. One transformer supplies either 120 volts and/or 240 volts (three-wire service).

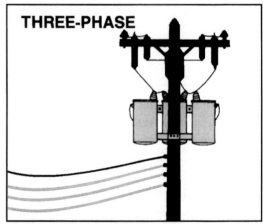

Figure 1-B-33. Three-phase power line. Either two or three transformers are required to supply three-phase service. There are four wires leading from transformers to the building, 120 and 240 volts can be made available.

One major electric utility reports infrequent use of 240 volts for three-phase service today, although it is still provided to farms. Instead, 120/208 volt service has been adopted for most 3-phase users. In selected installations of 120/208 volts, step-up transformers are necessary for ranges and clothes dryers to operate as they are designed (faster speed of heating at 240 volts).

Even if a three-phase power line passes a property, it does not necessarily mean a building has three-phase service. Single-phase service is supplied from three-phase lines.

To better understand the difference between single-phase and three-phase, think of one man driving a tent stake (Figure 1-B-34), then compare him with three men driving a tent stake (Figure 1-B-35). Three men, hitting one after the other, can deliver three times as many blows as one man working by himself. The same is true of three-phase power as compared to single-phase. Three-phase lines are used where large quantities of industrial power are needed. Most homes, farms, and small businesses do not need that much power, and are usually supplied with single-phase service.

Electric motors have to be built to operate with either single-phase power, or with three-phase power. They cannot be built to operate on both.

Figure 1-B-35. Three-phase service compares to three men driving the same stake.

Figure 1-B-34. Single-phase service is much like one man driving a tent stake.

11.
PHASE CONVERTERS

Electric power suppliers are required to provide single-phase power to all customers. This is usually adequate for residential and many commercial and light industrial loads. Since single-phase motors are available only up to 10 horsepower, equipment driven by larger electric motors will always require three-phase power. For example, irrigation pumps in rural areas typically need 15 to 100 horsepower and would require three-phase electric power (Figure 1-B-36).

Single-phase motors are considerably more expensive than three-phase motors of the same horsepower rating so many equipment manufacturers provide three-phase motors on their equipment regardless of horsepower. Typical tools using three-phase motors can include lathes, drill presses, compressors, saws and wood working equipment.

Because three-phase power lines are considerably more expensive to build and maintain than single-phase power lines, electric power suppliers study each request for three-phase power carefully. If the equipment to be operated uses less energy than needed to justify the $20,000 to $50,000 or more per mile cost for three-phase service, the power company has three choices:

- Require the user to pay the unrecoverable cost.
- Refuse to supply the three phase power.
- Suggest a phase converter be used to supply three phase power.

A **phase converter** is a device powered by single-phase service that produces three-phase service to operate electric motors and other electric equipment. This device permits the operation of three-phase motors from single-phase power. The size of the load is limited by the capacity of the single-phase power line, but loads of 100 horsepower (and over) are being successfully served with this type of equipment.

There are three types of phase converters commonly used today.

- **Electronic** - used on small loads up to 5 hp.
- **Static** - used on individual motor loads up to the capacity of the single-phase line (100 horsepower and up).
- **Rotary** - used on loads of single or multiple motors up to the capacity of the single-phase line (100 horsepower and up).

Some **variable speed drives** are employed as phase converters although their primary use is to provide special motor operating characteristics such as changing motor speeds, slow starting, slow stopping and other uses in industrial plants to synchronize production line processes.

Photo: Ivan L. Winsett

Figure 1-B-36. A phase converter can be used to convert single-phase power to three-phase power for motors in areas where it would otherwise be impractical such as rural application.

Understanding How Electrical Energy Is Used

Electric energy has so many uses that it is difficult to imagine being without it. In developed countries, it provides power for lighting, heating, cooling, refrigeration, manufacturing, entertainment, communications and a host of computers and other electronic equipment.

Electric energy supports improved **quality of life** in the home, community, and workplace. It provides a way to use the necessities and conveniences available. It also affords users many comforts, conveniences, and productivity that would not otherwise be possible.

As a result of the study of the material in this section, **the reader will be made more aware of the varied uses and importance of electricity in almost every area of daily life (see Table V). In addition, the reader will be able to define at least eight reasons why electric energy is used as well as identify and explain the terms and phrases listed in the "Important Terms and Phrases" box** found on this page.

The significance of electric energy to our families, our communities, our country and its economy, and our quality of life is discussed under the headings:

A. Important Uses of Electric Energy.
B. Why Electric Energy is Used.
C. Energy Management by Consumers.
D. Electric Safety–Home, Workplace, Community.
E. Emerging Technology in Consumer Electric Products.

Important Terms and Phrases

- Abundant, reliable, affordable, safe power
- Computer address
- CSA mark
- Electric energy consumers
- Electric safety
- Energy management
- Energy management priorities
- ETL certification mark
- EV (electric vehicle)
- Industrial uses
- Infrared heat sensors
- Miniaturization
- Motion/sound sensors
- Photoelectric cells
- Programmable central switching
- Quality of life
- Reason for dependency on electricity
- Remote controls
- Residential uses
- Robots
- Standard of living
- Timer controls
- UL certification mark
- User choices and behaviors
- Voice recognition
- Voice synthesis

TABLE V.
Classifications of Electric Power Applications

Type of Equipment	Typical Uses, Examples

Heating equipment (typically more energy intensive)	Space heating
	Major appliances (water heating, range)
	Portable appliances
	    Food processing (microwave oven, toaster, coffee maker, etc.)
	    Grooming (iron, curling iron, heating pad)

Motor driven equipment (typically uses less electricity per hour of use)	Major appliances (clothes washer, refrigerator, freezer) Portables Food preparation (blenders, food mixers, food processors, etc.)

Combination heating and motor driven	Forced air space heating
	Major appliances (clothes dryer, dishwasher)
	Portables
	    Food preparation (bread maker)
	    Grooming (hair dryer)

Lighting (typically uses less electricity per hour or use)	Incandescent, fluorescent, halogen, mercury vapor, sodium vapor, etc.
Controls	Manual operation of devices, automatic timers, thermostats, and a wide array of other controls

A. Important Uses of Electric Energy

Power suppliers classify **electric energy consumers** (users) according to types of applications or energy usage (Table IV and V). These uses will be discussed under the following headings:

1. Residential Uses of Electricity.
2. Commercial Uses of Electricity.
3. Industrial Uses of Electricity.
4. Street, Highway and Recreational Lighting Uses of Electricity.
5. Agricultural Uses of Electricity.

Within each of these categories, end uses of electrical energy power include:

- Lighting
- Motor-driven equipment
- Heating equipment
- Electronic equipment including computers and entertainment
- Combinations of two or more of the above.

1. RESIDENTIAL USES OF ELECTRICITY

Consumers are most familiar with electricity through its use in their homes. The most common association has come with use of lighting, water heating, appliances, televisions, VCR's, computers and the use of small tools. In addition, consumers seem to take for granted their homes will be equipped with heating and air conditioning (Figure 2-A-1). All these comfort and convenience items use electricity. Even though some homes use gas heating, cooking and clothes drying equipment, they are still dependent on electricity to ignite gas burners, to power fans to push warm or cool air through duct systems and to power motors, clocks, timers, controls, and lights on gas equipment.

At some time, almost everybody has experienced the loss of electricity due to storms or downed power lines. We realize just how much our daily lives are associated

Figure 2-A-1. Residential consumption is electric energy used in homes and residences. Heating, cooling, and water heating use the bulk of electric energy in the home. However, the oldest and most common use is for lighting.

with this source of power. Power outages lead to reduced productivity, comfort, convenience, efficiency, and safety. Unfortunately, power outages in some cities have been associated with bizarre vandalism, violence, burglaries, and other out-of-control behaviors.

2.
COMMERCIAL USES OF ELECTRICITY

Electricity is a critical link in delivery of services in every community. Commercial uses of electricity includes the lighting and power needs of huge shopping centers and malls. Additional places of use include; the critical needs of hospitals, the offices where we work, the restaurants where we eat, the hotels and banks we use, and the service stations where we get gasoline for our cars, (Figure 2-A-2). Other users are the movie

Figure 2-A-3. Electricity is used by industrial customers who manufacture one or more materials by processing or altering raw materials such as agricultural products, chemicals or metals to produce the products we all use.

houses we enjoy, the churches where we worship, the schools where we learn and all those places that contribute to our quality of life.

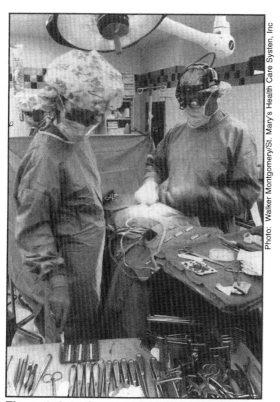

Figure 2-A-2. Commercial uses of electric energy include, among others; shopping centers, offices, restaurants, hotels, service stations, banks, movie theaters, churches and hospitals.

3.
INDUSTRIAL USES OF ELECTRICITY

Every business that produces and manufactures the items which consumers use on a daily basis must have electricity to power the machinery. Electricity is essential for cost effective manufacturing of parts and aids in assembling the finished products available in our consumer driven economy. The jobs of all employees in manufacturing facilities, both large and small, are affected by the need for electricity. Since places of employment **must have electricity** to operate, *jobs are dependent on an abundant, safe, affordable, and reliable supply of power* (Figure 2-A-3).

Figure 2-A-4. Street and highway lighting was one of the earliest uses of electrical energy. This lighting is used primarily to improve traffic safety and to aid in the protection from crime.

4.
STREET, HIGHWAY, PARKING, AND RECREATIONAL LIGHTING USES OF ELECTRICITY

Street and highway lighting was one of the earliest uses of electricity. In today's world, this use of electricity has grown to proportions never envisioned by governments and municipalities just a few decades ago. Research statistics reflect the fact that well lighted streets, parking lots, and communities are safer environments that reduce crime and improve traffic safety as well (Figure 2-A-4).

The emphasis on recreation and physical fitness has resulted in a fast growing use of electricity for lighting. Ballfields, tennis courts, stadiums, parks, resorts, and trails must now be lighted at night to accommodate the ever increasing numbers of users who work during the day and want to use recreational facilities during evening hours.

A passenger on an airliner, as well as distant astronauts, passing over any major American city can easily look down and see an example of this enormous use of electricity.

5.
AGRICULTURAL USES OF ELECTRICITY

Much of the use of electricity for agricultural applications is included in other areas we have discussed. For example, a farm or ranch may include a residence, a processing facility and a small shop or a feed mill. In addition, poultry and dairy operations make use of automated equipment dependent on electrical power (Figure 2-A-5).

Figure 2-A-5. Automated equipment in dairy and poultry operations are dependent on electrical power.

B. Why Electric Energy is Used

The everyday lifestyle and the **standard of living** of most people in the developed countries of the world depend in large part on the **availability of abundant, reliable, affordable, safe electric service**. To support today's quality of life, lights are expected to illuminate, and motors are expected to run at the touch of a switch - day or night, workday, weekend or holiday (Figure 2-B-1).

It is easy to understand why a disruption of electric service can cause work and play to come to a standstill. Downtime can be counterproductive. In some cases, a few individuals choose bizarre, disruptive, violent behavior when electric power is interrupted for a period of time.

The **reasons for the dependency on electricity** include:

Quality of electric lighting and other equipment. There is no other readily available and inexpensive substitute for the quality lighting achieved with electrical powered devices and lighting fixtures.

Safety of electric energy. Although improper use of electricity can result in injury, death, and loss of property, properly handled electrical energy is used by millions of people every day, mostly with no accidents or injury.

Convenience. Electric energy is as convenient as turning on a switch.

Cleanliness. Electric energy is odorless and colorless at its point of use. There is no cleaner form of energy available than electricity. Electrically powered appliances (water heater, clothes washer, dishwasher, refrigerator, etc.) support sanitation and personal health standards which contribute to wellness, longevity, and the highest standard of living known to man.

Photo: David Gentry, Warren RECC

Figure 2-B-1. A kitchen in today's home provides an opportunity to benefit from all the pluses of electric energy through quality lighting, convenience, cleanliness, safety, dependability and economy.

Dependable. The training of personnel and preparedness of power suppliers have made electric energy very dependable in all but extreme conditions. This reliability of power is a hallmark of the electric industry. Reliability of power is facilitated by grids made up of agreements among utility companies to share power during approaching outages, peak periods, and black outs.

Consumers, on occasion, cause disruption of power when automobile accidents disable utility poles and equipment or other consumers initiate damage to utility equipment.

Efficiency. Equipment operated by electricity is highly efficient compared to other sources of energy. Manufacturers and utility companies continue to improve efficiencies of products and services.

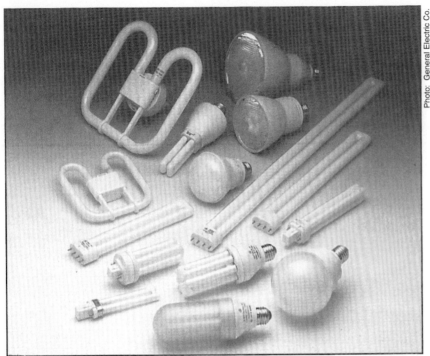

Figure 2-B-2. Economical electricity powers technological innovations which afford long life, improved quality output, greater efficiency, and enhanced safety.

Economy. Electric energy is the most affordable type of energy available for the diverse uses made of its capabilities. The diversity of electrical applications includes numerous technological innovations affording users more energy effective lighting devices (Figure 2-B-2).

Productivity. Electricity powers the tools and equipment many of us use in our employment as well as at home (Figure 2-B-3). We are more productive because of the functions these machines provide us where we work. Specifically, electricity provides the power to increase worker output per hour. Electricity literally powered the Industrial Revolution and its attendant development and widespread use of electric equipment for homes, workplaces, and communities.

Advances in electrically powered computers, microprocessors, programmable logic controllers (PLC), and variable speed drives have boosted productivity through precise control.

Figure 2-B-3. Electrically powered tools are a necessary part of many jobs in the construction industry.

C. Energy Management by Consumers

Since the 1973-74 oil embargo by the Organization of Petroleum Exporting Companies (OPEC), **energy management** has been an issue of varying intensity with American consumers and policymakers. There are a number of reasons for wise energy management by consumers whether they be at home, in the community or workplace (Figure 2-C-1). Two leading reasons are economics and environment. These reasons continue to influence:

- Research, development and design of equipment, appliances and vehicles.
- Consumer purchases of energy using products
- User behaviors regarding "on" time of appliances and thermostat settings of heating and air conditioning equipment.

Consumer energy management of electricity is discussed under the following headings:

1. Conservation vs. Management Strategies.
2. Energy Management Priorities.

1. CONSERVATION VS. MANAGEMENT STRATEGIES

Energy conservation programs in the 1970's stressed energy reductions which could have the side effects of reduction of quality of life and productivity. Often consumer/employee apathy and outright rejection resulted from unappealing programs.

Now, energy management programs promote decisions/strategies which seek to balance energy and economic reductions with user safety, health, productivity, stress, and comfort issues. In general, energy management strategies are more user friendly and encourage voluntary selection of appropriate priorities over severe conservation measures and forced energy cut-backs.

More recently, demand-side (consumer) energy management programs have been developed and promoted by electric utilities. This type of energy management program

---

Benefits of Wise Energy Management

- ❯ Lower monthly costs for energy use at home and in the workplace and community

- ❯ Increased use of more enviromentally responsible choices

- ❯ Increased social awareness and responsibility for the needs of successive generations of users

- ❯ Lower the current demand for fossil fuels--nonrenewable energy forms

- ❯ Stretch the use of existing fixed sum supply of nonrenewable fossil fuels (coal, oil and natural gas)

- ❯ "Buy time" to develop safe, abundant, reliable and affordable alternative energy resources and attendant technology necessary for a future energy conversion

- ❯ Lessen the vulnerability of U.S. households and economy to energy supplies from foreign sources

- ❯ Decrease the increasing balance of trade gap between our country and overseas suppliers of energy products

- ❯ Reinforcement of stewardship of resources as an ethical standard.

encourages users to carefully choose which energy uses will be reduced and when and by whom. Energy management is a proactive, cooperatively responsible strategy.

2.
ENERGY MANAGEMENT PRIORITIES

Frequently consumers and employees are confused about which appliances and equipment use the most kilowatts of energy in a day's time or perhaps over a month. Without an understanding of the comparative amount of energy used by appliances and equipment, a consumer may reduce the "on" time of a particular electrical product with little or no impact on the electric bill. Other products use many times the amount of electrical energy of others over a year or month, or even a day or single hour of use.

If consumers of electricity are to effectively manage usage, then certain **energy management priorities** must be recognized and addressed. The following list cites electric energy uses (from the highest to the least) that would be typical of many families, although not all:
- Central heating
- Central cooling
- Water heating
- Refrigerator
- Freezer
- All lighting
- Clothes dryer
- Electric range
- Clothes washer (higher on the list if energy for hot water is included)
- Dishwasher
- Portables that heat (hairdryer, etc.)
- Motor driven portables (power saws, vacuum cleaners, etc.)
- Television
- Computer systems.
- Sound equipment (radio, stereo, etc.)

Photo: James W. Strawser

Figure 2-C-2. Maintaining heat or cooling at reasonable levels can help control energy expenses.

There are a number of exceptions the reader should consider and take into account when looking at this list of priorities. For example, while families in cooler climates might use little or no central cooling, Florida families would sustain minimal central heating costs. The use of a range and a refrigerator would be diminished if a family eats out often. Small households could have very little laundry and thus have little washer/dryer usage. Users must evaluate and consider their own particular needs and determine adjustments that can be implemented for maximum energy savings.

On balance, electric heating and cooling equipment is generally the largest user of electricity. Programmable thermostats for heating and cooling can be set up by a homeowner to come "on" during waking hours and "coast" during sleeping hours. Well thought out and informed purchases and use of electrical equipment can reduce the "on" time and is highly recommended to save energy (Figure 2-C-2).

In summary, effective energy management depends on an individual **user's choices and behavior**s at the point of selection and during use of electrical equipment.

D. Electrical Safety - Home, Workplace, Community

Electric energy powers a wide array of equipment and motor driven items that most people believe are necessary to our daily lives in our homes, at our workplaces and in the communities in which we live. We often take this power for granted and assume that electric energy serves us with no responsibility expected from the user... other than paying the monthly bill. We tend to overlook the second, but equally important user responsibility–**electrical safety**–an issue with any energy resource.

While electricity is a servant, this friend should be treated with knowledge and respect. Used responsibly, safe use of electricity provides a large list of benefits to individuals and organizations (see Chapter 2, A. Important Uses of Electric Energy).

Since the safe use of electric power is a concern of everyone, it is important that we know about other organizations concerned with its safe use. It is also important that we recognize some of the typical electrical hazards we can all come in contact with. These areas are discussed under the following headings:

1. Electrical Codes.
2. Recognized Product Safety Testing.
3. The National Electrical Safety Foundation.
4. Recognizing Typical Electrical Hazards.
5. Addressing Electrical Safety Problem.

1. ELECTRICAL CODES

The need for standards for the safe and correct application of electrical installations for a variety of uses was recognized shortly after Thomas Edison's invention of light bulb technology and the founding of the first utility system in New York City in the late 1800's. The first electric code document was developed in 1897 as a united effort of fire and insurance organizations and groups recognizing the need for safety standards.

In 1911 The National Fire Protection Association began its sponsorship of the *National Electric Code®*. This sponsorship continues today and a revised *Code* is printed every three years. This code addresses: (1) changing power needs, (2) safe user practices of new technology (Figure 2-D-1), and (3) increased energy use in the home, workplace

Photo: David Gentry, Warren RECC

Figure 2-D-1. All wiring in new and remodeled construction must comply with local and national building regulations.

and community. The *National Electric Code*® is the minimum standard for electric wiring practices and materials used nationwide.

Electric utilities, along with electrical inspectors, often address local or regional safety concerns by adding specific standards for inspection of electric wiring in new and remodeled buildings and other installations (Figure 2-D-1). Building owners should always request the results of the electrical inspection. Determining electric code compliance assures a minimum level of safety for occupants of the premises.

2.
RECOGNIZED PRODUCT SAFETY TESTING

Three major product safety testing organizations are: Underwriters Laboratories Inc., Intertek Testing Service NA Inc., and Canadian Standards Association. The meaning of their marks, usually found on the bottom or back of a product, are discussed here.

Underwriters Laboratories Inc. (UL). A century old independent not for profit testing and certification organization that evaluates products, materials and systems in the interest of safety. A **UL certification mark** on a product or device means representative samples have been tested and evaluated with reference to regularly updated safety standards for electric shock, fire and related safety hazards (Figure 2-D-2).

Intertek Testing Services. (ETL). The **ETL Listed certification mark** found on a product is a sign that two activities have happened. The product has been tested to assure it conforms to an appropriate safety standard and the manufacturer's factory has been (and continues to be) inspected to assure ongoing compliance. The ETL Listed Mark appears on electrical and gas-powered products ranging from computers to industrial equipment (Figure 2-D-2).

Canadian Standards Association. (CSA). The **CSA mark** is registered in Canada and other countries. When it is displayed, it tells the purchaser and users that their products

Figure 2-D-2. The "UL", "ETL" and "CSA" marks indicate that samples of a product have passed tests for safety applicable to U.S. standards.

meet the applicable requirements of a standard. Only products that are tested and certified by CSA (or by an organization qualified by CSA) are allowed to bear the CSA mark (Figure 2-D-2).

Safety and regulatory organizations in the U.S. and other countries recognize the reports, test data and listing information concerning products bearing one or more of these marks.

3.
THE NATIONAL ELECTRICAL SAFETY FOUNDATION

The National Electrical Safety Foundation (NESF) was chartered in 1994 to promote electrical safety in the home, school and workplace. The focus of NESF is to prevent personal injury and the loss of life and property. NESF reports 3 people die each day from dangerous contact with electricity in the U.S. Proactive prevention of electrical hazards is as important as identifying existing electrical hazards and following through with safe problem-solving practices (Figure 2-D-3).

The NESF is collaborating with trade/professional organizations including the Canadian Standards Association, International Association of Electrical Inspectors, International Brotherhood of Electrical Workers, Association of Home Appliance Manufacturers, National Fire Protection Association, National

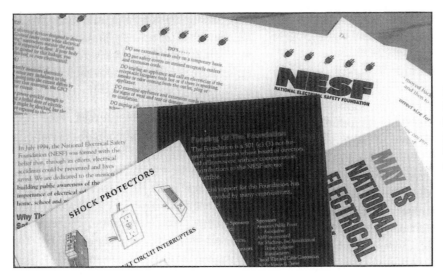

Figure 2-D-3. The focus of the National Electrical Safety Foundation is to prevent injury and death through the proper use of electricity.

Electrical Contractors, National Electric Manufacturers Association, investor owned electric utilities (Edison Electric Institute), municipal utilities (American Public Power Association), rural electric utilities (National Rural Electric Cooperative Association), Underwriters Laboratories; manufacturers and retailers and the U.S. Consumer Product Safety Commission in the cause of upgrading electrical safety.

4.
RECOGNIZING TYPICAL ELECTRICAL HAZARDS

Electrical safety hazards can result from storms or other non-human reasons, but most safety hazards result from dangerous choices and behaviors by consumers of electricity. These human behaviors may not always create unsafe electrical conditions. However, any electrical hazard is a constant risk and threat to personal health, human life, and preservation of property. These **risks pose the problem of when–not if–a disaster will occur**. Several typical hazards are identified and discussed as follows:

- **Misuse of electrical equipment.** Adults and children should be taught to use electric equipment carefully. Children should be taught the danger of climbing antennas, trees, buildings or poles that could touch power lines. In addition, everyone should have an awareness of the danger associated with the vandalism of

electric equipment. Besides destroying equipment belonging to the utility company and driving up costs to consumers, vandalism can result in immediate death from the instant flow of electrical current.

- **Improper connecting and disconnecting of electric plugs.** All equipment should be turned off before it is connected or disconnected. Always use dry hands and then firmly grasp the plug to insert (connect) or remove (disconnect) the plug from an electric outlet (Figure 2-D-4). Always stand on a dry surface and never touch the prongs on the plug in this process. Do not pull or yank on the cord-this can loosen wiring inside the plug which may later cause arcing, sparks and overheating.

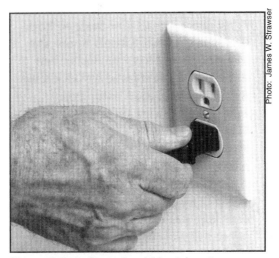

Photo: James W. Strawser

Figure 2-D-4. Care should be taken to properly connect or disconnect plugs to or from an electrical outlet.

- **Improper positioning of electric cords.** Too often the insulation around conductors in electric cords under rugs, across doorways, or near heat, will become worn, frayed, cracked and damaged by traffic and heat. Damaged wires can cause shorting or sparking which could be the cause of a fire. The bare wires, if contacting each other, could cause the breaker to trip at the panelboard. Exposed wires also provide dangerous paths through any conductor, including the human body (shock or electrocution). Short circuits and electrical faults are defined in the glossary.

- **Placement of electric equipment near water.** In general, electric equipment and conductors should be kept away from water. Electric current can quickly follow unsafe paths in bath water, dishwater, laundry water, ground/surface water, pools, fountains and rainwater. A GFCI (Ground Fault Circuit Interrupter) should be installed in any location where a power receptacle is near equipment using water, i.e. pumps, outdoor pools, or other potentially unsafe locations. When properly installed, a GFCI will turn off the power immediately before unsafe amounts of current begin flowing through people.[5]

- **Incorrect sizing of electric fuses.** When an electric fuse is replaced, care should be taken to ensure that the replacement fuse **is not** larger than the one being replaced. The appropriate fuse would be one that matches the wire size (conductor) and electrical current allowed on that circuit. See Table III on page 19 for more information on wiring size requirements. Never use fuses with larger capacities than recommended because overfusing can permit unsafe amounts of current flow through the circuits. Never place a penny or a wad of aluminum foil behind a disabled fuse. This is an unsafe practice which will provide **no** protection from excess current and overheating of wiring.

Overfusing, the use of pennies and foil are dangerous practices. Sooner or later, at a time when too many electrical appliances are plugged into such a circuit, a fire will be the likely result.

- **Improperly sized conductors.** The conductor (wire) must be large enough (diameter) and be made of material that will match the amount of electric current needed (see Table III). Most conductors will be made of copper or aluminum. Conductors that are too small or made from poorly conducting material offer resistance to excessive amounts of electrical flow. The consequences of this excessive resistance is overheating of the conductor and perhaps a fire in nearby materials.

- **Arcing and sparks from electrical equipment.** Arcing occurs if the switch is "on" during either the plugging or unplugging of an appliance or tool. The "on" switch "demands" electric current which leaps from the outlet when the prongs of the plug are a short distance from the outlet. Arcing can shock humans if their fingers or other conductors engage the electric arc.

To prevent this arcing menace, simply turn all equipment to the "off" position before connecting or disconnecting electrical equipment. Unfortunately, some appliances such as popcorn poppers do not have a switch.

- **Modification of grounding prongs.** Many of today's 120 volt electric products or extension cords have a 3-prong plug. One of the three prongs is a grounding prong designed to conduct dangerous short circuits from appliances or metal surfaces safely to the main panelboard to trip the circuit off if the circuit is wired properly. Newer plug receptacles are designed to accept 2-flat blade polarized prongs and a round, grounding prong. Older plug receptacles do not provide an inlet for the ground prong. The grounding prong **should never be removed** to make a 3-prong plut fit a 2-prong receptacle. A better

[5]Any appliance can use an excessive amount of electricity through a GFCI and the GFCI will not trip. The only time a GFCI trips is when the current going into the appliance is different from that returning. It is not an overcurrent device.

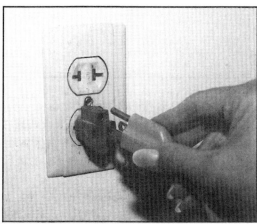

Figure 2-D-5. A receptacle adapter can be used only with grounded receptacles and the grounding wire or loop of the adapter must be attached to the cover plate mounting screw. The mounting screw connects the adapter to the grounded metal electrical box.

practice is an inexpensive grounding adapter properly connected to the receptacle (Figure 2-D-5). Use it **only if** the receptacle connects to a grounding wire or grounded electrical box.

- **Use of electric equipment near wet, damp areas.** Avoid working and playing with electric equipment on damp lawns, patios, decks and other outdoor or wet areas. Remember, both water and damp earth are conductors of electricity. Electrical current can instantly and dangerously affect users of electric appliances in damp areas.

- **Damage to electric wiring during fires.** When a building fire occurs, insulation of electric conductors can be damaged. When this occurs, electric current can flow through any adjacent conductor, including water sprayed to control the fire! The safe reaction is to call the electric utility to turn off all power to the building. Then stay away from the fire site until all power has been disconnected.

- **Temporary wiring hazard.** Do not allow temporary wiring or connection to become a permanent hazard. Extension cords (especially to window air conditioners), spliced supply cords to mobile homes left lying on the ground, and temporary connections to new equipment are examples of this hazard.

5.
ADDRESSING ELECTRIC SAFETY PROBLEMS

When it has been determined that an electric hazard exists, the first step is to identify the problem before attempting any corrective action. Suggested practices to follow after an **electrical safety hazard** has been identified include the following:

1. *Correct or remove the hazard as soon as possible.*
   Prevent tragedy before it occurs. Contact an electrician, local utility company or other professional if more information or help is needed (Figure 2-D-6). **Never** ignore an electric hazard. It is not a question of "if" dangerous results will occur. **Rather**, the risk involves "when" loss of property, personal injury and/or death occurs.

2. *If you are shocked by electricity, contact the appropriate health professionals immediately.*

3. *If someone else is shocked, be very careful to avoid liability by knowing the appropriate first aid measures to take.*

Figure 2-D-6. The utility company should be contacted when power lines are down as a result of a storm or accident.

E. Emerging Technology in Consumer Electric Products

In recent times, projects such as the *Smart House,* sponsored by the National Association of Home Builders, the *Bright Homes,* showcased by Indianapolis Power and Light Company, and the *Good Cents* program, have done much to stimulate consumer awareness of emerging electrical technology (Figure 2-E-1).

Exciting new products are emerging in the consumer marketplace. Others remain in the research and developmental and have not been accepted by consumers. Only time will tell if the technological innovations discussed in this chapter will receive wide scale acceptance by the public. Much of this technology is particularly useful to physically challenged consumers and employees and those concerned with energy conservation and costs.

This discussion is not all-inclusive or extensive. It is designed to make the reader aware of selected innovations, new products, and developing technologies.

These innovations are discussed under the headings:

1. Electric Controls.
2. Miniaturization of Components.
3. Voice Recognition Devices.
4. Voice Synthesis Products.
5. Increased Use of Robots.
6. Electric Vehicles (EV)

1. ELECTRIC CONTROLS

All of us are familiar, to some extent, with electric controls development whether it be through the use of the remote control devices or timed thermostats on heating and cooling equipment. Some of the areas where controls are becoming more important in our daily lives are:

- **Remote Controls.** Hand held remote controls can be used to turn almost any electric equipment "on" or "off." In fact, each electric outlet in the home could have a **computer address** to enable that outlet to be controlled from one or more central or portable control stations in or outside the home. It is possible to turn the electric range "off" while in transit to one's workplace or to turn on selected lighting from a vehicle as the owner nears home.

- **Photoelectric Cells.** Although photoelectric switches have been widely used to turn on street lights, parking lot lights, highway interchange lighting and individual home security lights, this technology has been underused in interior spaces. Portable photoelectric

Figure 2-E-1. Many of the inovations that consumers will take for granted in the future can be seen now in homes designed to incorporate technological advances. Some are designed to conserve electricity, others to improve the quality of life.

Photo: Georgia Power Co.

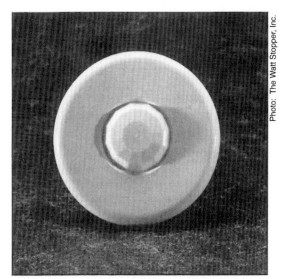

Figure 2-E-2. Electric controls have advantages to consumers that can save energy and money such as this ceiling mounted passive infrared occupancy sensor.

cells can be plugged into receptacles near windows, for example, to automatically activate lighting devices inside a home near sunset for safety and security purposes.

- **Timer Controls.** Although timer controls are quite affordable, there has been an underuse of the capabilities of this convenient product. Permanent and portable timer controls can turn electric devices on/off automatically. Users determine the on and off times. Timer controlled water and space heating equipment can effectively manage to reduce peak energy usage with a resultant cost savings to the consumer. This technology is greatly underused given the improved quality of life and financial savings that can be realized.

- **Motion/Sound Sensors.** These detectors can automatically turn electric equipment on/off. For example, room lighting will automatically be turned "off" if there is no motion for a programmed period of time. In other cases, lights will be turned "on" by car

or body movement or other form of motion detected. This available technology is not fully utilized at the present time.

- **Infrared Heat Sensors.** These sensors will detect the body heat from human bodies (98.6°) and energize heating or air conditioning equipment for room usage (Figure 2-E-2). When the heat source leaves the room, the equipment will be automatically turned off or reduced. This type of sensor can be especially useful to physically impaired consumers.

- **Master, Programmable Central Switching.** This type of control has been available for over twenty years. Unfortunately, its safety, security and convenience merits have not been realized among the majority of consumers. One or two central switching sites enable the user to turn on/off/dim any and all light sources inside or outside a building from one convenient, secure point.

2.
MINIATURIZATION OF COMPONENTS

Although miniaturization is not a new concept, continued developments have brought about decreases in the size of controls and equipment while the complexity of the wiring and circuitry has increased. Touch pad controls on microwave ovens and calculators are both examples of everyday items that have been improved by these changes. A great part of the success and advancements in this area are positive by-products of the U.S. aeronautical space research program.

Consequences of miniaturization are great increases in compact design, improved performance, enhanced reliability, longer life and the versatility of products at an affordable price enjoyed by all of us.

3.
VOICE RECOGNITION DEVICES

At some time in the not-too-distant future, consumers may be able to speak into their home security system to open locked doors rather than use a key. The human voice is unique, much like a fingerprint, and in the future developing technology may take advantage of this distinctness to the advantage of consumers.

4.
VOICE SYNTHESIS PRODUCTS

While a few automobiles, vending machines and other assorted devices speak to their users, this emerging technology lacks widespread use. One appliance manufacturer offers voice synthesis features which provide messages to blind consumers. Vehicles with this feature remind drivers that keys have been left in the ignition, that the lights have been left on, or that a door is ajar. Vending machines may remind us to insert certain coins or prompt us to make a selection after the money has been deposited in the machine.

Figure 2-E-3. Electric vehicles are recharged via cords and paddles specially designed for the purpose. Recharginng stations may become a common sight in the future.

Voice synthesis coupled with computers could support a variety of security measures, consumer services, and benefits for the physically challenged.

5.
INCREASED USE OF ROBOTS

Robots are used with increased frequency in industrial applications, Robots are already used by industry to pin stripe cars after another robot has painted the vehicles. However, little use has been made of their capabilities in homes. If costs, versatility and other operations are modified, perhaps robots will mow our lawns, vacuum the floors and do a multitude of other jobs around the house.

6.
ELECTRIC VEHICLES (EV)

Battery powered electric automobiles have been extensively tested over the past twenty years. Non-polluting automobiles are now mandated by legislative deadlines in some states. Electric powered buses are being used in some cities and the use of electrically powered vehicles by postal and police is becoming more widespread. Speed and mileage potential of EVs have increased as a result of research in recent years.

Batteries for EVs can be recharged via a cord with a paddle inserted in compatible electrical receptacles (Figure 2-E-3). Widespread use of the electric car would use significant amounts of electricity. The electric energy to recharge the batteries will come from the electrical power system.

EVs operate quietly and require no ignition of a motor before engaging the drive function. Only a few electric powered vehicles are available at this time, but all U.S. manufacturers are developing electric cars. They are in various stages of testing at this time.

How Electrical Energy Is Generated

Chapter 1 informed readers that electric current is the flow of electrons in a circuit. Starting electron flow is known as "generating" electric power. In fact, approximately one-fourth of all energy consumed in the U.S. is used in producing electric energy. Other forms of energy include oil, nuclear and natural gas.

Electric energy is not created; it is converted (generated) from other forms of energy. In fact, one of the basic laws of science is that energy is neither created nor destroyed. Thus, it is converted from one form to another--from falling water, or burning coal, solar energy, or splitting the nucleus of uranium to produce electricity.

The purpose of this section is to provide the reader with a basic understanding of how electric energy is mechanically generated. As a result of studying this section **the reader should be able to identify the different methods of driving mechanical generators. In addition, he/she will be able to identify and describe the related terms listed in the "Important Terms and Phrases" box** found on this page.

How electric energy is generated is discussed under the following headings:

A. Renewable and Nonrenewable Resources.
B. How Mechanical Generators are Driven.
C. Considerations for the Future.

Important Terms and Phrases

Active solar	Nuclear reactor
Breeder reactor	Oil
Coal	Passive solar
Criteria for energy alternatives	Primary system
Fissionable fuel	Renewable resources
Fixed sum fuels	Secondary system
Fossil fuels	Seismic faults
Fuel cell	Solar voltaic
Fusion	Solid waste
Geothermal energy	Steam power
Hydroelectric generators	Tidal
Mechanical generators	Turbine wheel
Natural gas	Water power
Nonrenewable resources	Wind
Nuclear fission	Wood

A. Renewable and Nonrenewable Resources

Each past, present, and future energy form can be classified as a **renewable** or **nonrenewable** resource. **Renewable resources** are valuable for their supply and does not diminish after an hour's, a day's, or even a year's use. Solar, tides, geothermal, wind power, solid wastes, water, and perhaps fast growing woods are renewable resources.

Nonrenewable resources include all fossil fuels (coal, oil, and natural gas), non-sustainable woods and nuclear (uranium) energy supplies (Figure 3-A-1). Users of those **fixed sum fuels** face eventually exhaustable, increasingly difficult to secure expensive supplies if increasing demand by consumers and other users continues.

Innovations and inventions of the early 20th century resulted in a nationwide energy conversion from a dependency on manpower, animal power, wind, and water power to the use of fossil fuels. As fossil fuels decline in availability, Americans are likely to sustain another energy conversion to emerging energy alternatives in the future.

When a ton of coal, a gallon of oil/gasoline or a cubic foot of natural gas is used, it can never be reused again. Data indicates that during the twentieth century the U.S. demand for energy has doubled in as few as every seven years! These increases continue to deplete the fixed sum of remaining fossil fuels at an accelerated rate.

The consequence of energy use decisions and actions by individual consumers and employers, tends to drive the price of energy upward. Heavy demand also increases imports into the U.S. and dependence on supplies from other nations. Increased fuel demands and a decreased supply of nonrenewable resources will determine the price that must be paid in the future. Wise energy management will help curb higher costs and extend the availability of nonrenewable resources.

Figure 3-A-1. Coal is a non-renewable resource that is used in large quanities to produce electric power. In 1996 over 874 million tons were used by electric producers nationwide.

B. How Mechanical Generators are Driven

Electric energy is mechanically generated by passing coils (spinning or turning) of wire through a magnetic field. Both direct current (DC) and alternating current (AC), discussed earlier in this book, may be generated by **mechanical generators.**

Three methods of driving mechanical generators are described and explained as follows:

1. Water Power.
2. Steam Power.
3. Other Energy Forms.

1.
WATER POWER
(Renewable Energy Resource)

Water flowing from a higher to a lower level can be used to turn a generator (Figure 3-B-1). This action is called **hydroelectric generation**. This type of electric energy generation developed very rapidly after 1910. Approximately 11% of electricity generated in the U.S. in 1995 was hydroelectric. By now, the best hydroelectric generation sites in North America have been developed. However, there are still many good potential sites in underdeveloped countries.

Large streams and rivers with one or more dams create manmade reservoirs which provide for controlled flows of water (Figure 3-B-2). Water power uses the principal of the ancient water wheel to turn a generator. The "water wheel" is constructed of metal and is called a "**turbine wheel**." The electric generator is attached to the turbine shaft. Water flows against the turbine blades causing both the turbine and the generator (rotor) to turn. The coils of wire are turned through a magnetic field, consequently, electrons flow along suitable conductors from the generating site.

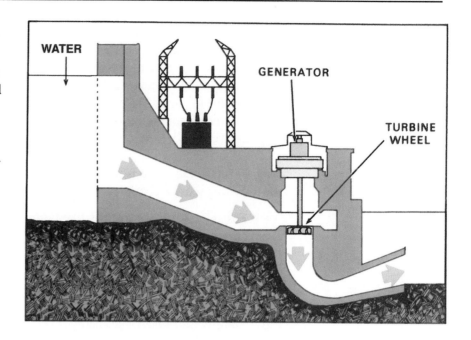

Photo: U.S. Department of the Interior, Bureau of Reclamation

Figure 3-B-2. A hydroelectric generating plant.

2.
STEAM POWER

The vast majority of all electricity is produced by **steampowered generators** (sometimes referred to as thermal-powered generators). Water is heated in a boiler converting the liquid to steam. Steam, at high temperatures and pressures, causes the blades or fins of the turbine to rotate. The electric generator (rotor), being connected to the turbine, also turns (Figure 3-B-3).

Heat for producing steam is supplied from the following sources of energy:

a. Fossil Fuels: nonrenewable, fixed sum energy resources
b. Nuclear Fission: nonrenewable, fixed sum energy resource
c. Geothermal: renewable energy resource
d. Solid Waste: renewable energy resource
e. Wood: a renewable energy source; slow growing trees are not

There are other sources of energy for development as alternative energy forms. Criteria for acceptability of energy alternatives and energy conversion are topics at the end of this Chapter.

a. Fossil Fuels
(Nonrenewable energy resource)

Fossil fuels have fixed sum supplies and are nonrenewable energy either mined or piped from the earth. Fossil fuels used in producing steam represent the three states of matter and are as follows:

- Coal (solid state)
- Oil (liquid state)
- Natural gas (gaseous state)

Coal. Coal is the fuel most commonly used to make steam for generating electric energy (51% in 1995). At the present time, there is still a rather abundant supply of coal and it can be obtained at a relatively low cost. However in the future, users will face dwindling quantities and the quality will likely decline. The remaining U.S. coal that is abundant is softer than the most desirable harder varieties.

Today's supplies from the Appalachian and Rocky Mountains are softer and more polluting and have less heat producing potential than previous coal used for this purpose. Future use of coal from dwindling, more costly supplies will more likely be extracted from strip mines than from almost "mined out" deep mines. Strip mines have serious land and environmental consequences and/or costly land reclamation requirements in some states. The softer coal remaining in Appalachia has a high sulfur content which must be "scrubbed" from emissions in smokestacks of coal-fired generators in order to comply with federal requirements enforced by the Environmental Protection Agency (EPA). This pollutant produces sulfur dioxide (SO_2). The Rocky Mountains have vast supplies of coal which yield nitrogen oxides rising in smokestacks of electric generators. These oxide emissions are costly to remove and are monitored by the EPA to assure air quality standards are maintained.

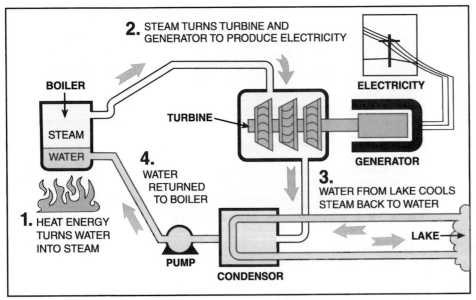

Figure 3-B-3. Principle of a steampowered generator.

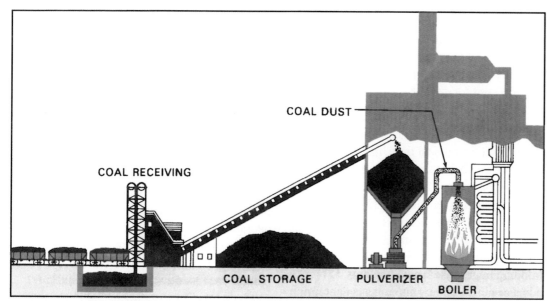

Figure 3-B-4. How steam is produced in a coal-burning generating plant. Steam is produced when the boiler of burning coal heats water. Steam under pressure turns a turbine in a magnetic field which generates electron flow.

While supplies of nonrenewable, harder, cleaner burning and greater heat producing coal are declining, the generation industry continues to turn to softer coal. Thus, future supplies pose more consequences from (1) lower heat producing efficiency per ton of coal, (2) increasing pollutants from generation, and (3) the increasing costs of technology to support air quality related standards for electric generation.

When available, a 90-day supply is usually maintained at most power generation plants. The coal is carried by conveyor belts from a coal pile into the plant. Massive pulverizers literally reduce the coal to dust. The coal dust is then blown into the firebox of the boiler under high pressure. It is burned almost instantly, heating water which produces steam for driving the turbine and generator. (Figure 3-B-4).

Oil. While oil is the most common fuel used to power transportation in developed nations, oil is also a fossil fuel for generating electricity at steam plants (Figure 3-B-5). In 1995, 2% of U.S. electricity was generated by oil-fired plants. Most of the oil used is of the heavy residual oil type. For many years, oil was preferred over coal as it was easier to handle and cleaner burning. Therefore, many plants were converted from coal to oil.

However most of these plants have been converted back to coal because of the faster increasing costs of oil and the uncertainty of an adequate supply. A large percentage of oil is imported into the U.S. while domestic wells supplying oil economically and in large supplies continues to decline. While Alaskan oil is generous in supply, much of it is so-called "sour" oil which is incompatible with U.S. refineries, and thus, is exported.

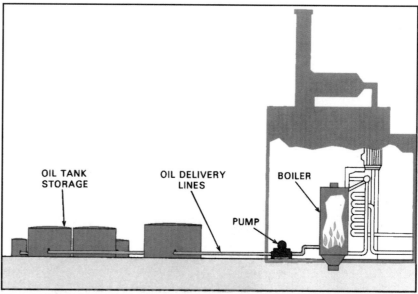

Figure 3-B-5. How steam is produced in a oil-burning generating plant where pressurized steam is used to generate flow of electrons.

Natural Gas. In years past, gas was supplied to fuel electric generators to provide energy for peaking needs. The percentage of electricity generated by plants using natural gas decreased for a number of years because of a variety of transport and government regulation factors.

However, at the present time, natural gas is becoming the fuel of choice for some utilities. The reasons are many and include: (1) use of existing technology, (2) low capital cost, (3) short license procedures, (4) ability to add small amounts of distributed generation[6] for growth needs without building huge and expensive plants, (5) advances in efficiency making gas generation more cost effective, and (6) lower environmental impacts. Much of this move towards gas has resulted from the impact of deregulation in the gas industry and the convergence (mergers) of gas and electric utilities into "energy companies".

Today approximately 15% of the electricity in the U.S. comes from natural gas-fired plants. Utilities can select a variety of fuel options based on availability and market prices.

[6]Distributed generation is essentially adding generation at substations to meet growth or peak load requirements and provide voltage support. It also reduces transmission losses.

Imports of natural gas may be likely in the future as the increased use of gas is needed for heating buildings and for industrial processes.

b. Nuclear Fission
(Nonrenewable uranium sources)

Splitting uranium atoms yields vast amounts of heat for generation of electricity in nuclear plants. There are scores of nuclear plants in operation in North America (Figure 3-B-6). Nuclear energy now generates about 20% of the nation's electricity. To date, nuclear plants use **fissionable fuel** (atom splitting) and, due to design, are not as likely to melt down as many people think. Steam turbines similar to those used with other fuels, are used in nuclear plants. Atom splitting (fission) within a **nuclear reactor** produces heat used to heat water which yields steam needed to turn the turbines to generate electricity.

The water is heated by the fission process in the uranium core area within the **nuclear reactor** and moves through a closed system

Photo: Georgia Power Company

Figure 3-B-6. A nuclear generating plant.

48 CHAPTER THREE

known as the **primary system** to the steam generator. This water is under pressure so that it will not boil and is circulated by a pump in this closed system (Figure 3-B-7). Water is also used to prevent the uranium fuel from becoming too hot and unsafe. A separate lower-pressure water system, known as the **secondary system**, is introduced to the turbine area. This water absorbs the heat from the primary system and, since the pressure in the secondary system is lower, it boils and turns to steam. Generally water is removed to produce a "dry" steam at much higher temperatures. This is piped to the turbine area to turn the turbine and generate electricity. The steam is circulated through a condenser immediately under the turbine to produce a vacuum creating tremendous power with the steam rushing past the turbine fins.

Once condensed, the dry steam is recycled and reheated. The cooling water flowing through the condenser is returned to the lake, river, or cooling tower.

There exists a second type of nuclear fission where atom splitting yields even more heat energy for electric generation than typical nuclear reactors. These **breeder reactors** create plutonium to enrich the uranium fuel core and are considered to be a source for producing weapons grade materials. Consequently, many American nuclear engineers, including former President Jimmy Carter, have effectively opposed this form of generation of electricity in the U.S. However, France, China, and India lead the world in breeder reactor use followed by much of Europe, Israel, and developing countries that do not have (or have limited) fossil fuel resources for generating affordable electricity.

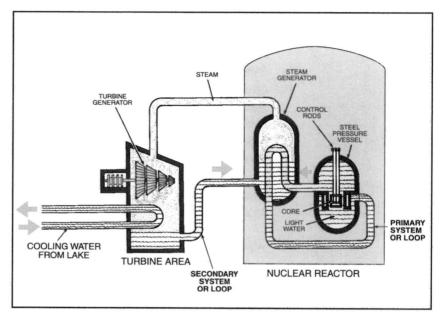

Figure 3-B-7. Cross section of a pressurized nuclear reactor (PWR) supplying heat for a steam generating plant.

c. Geothermal
(Renewable energy resource)

Geothermal energy is caused by normal **seismic faults** (breaks in the earth's crust). Water, seeping deep into the earth's surface to heated rocks, turns into steam which can then be used to operate steam-powered electric generators (Figure 3-B-8).

Figure 3-B-8. Geothermal steam can be harnessed to operate steam-powered electric generators.

Geothermal energy is limited to geographic locations where natural geysers occur. At times, one-third of the electricity supply for San Francisco, California, is supplied by electric generators powered by natural steam from geothermal sources nearby. Currently, electric generation is regional, thus dependent on geothermal steam "pockets" of energy. Hot steam and natural geysers, as useful as they are as natural steam sources which bypass the need for heating water, are available in few areas.

While currently not useful for electricity generation, geothermal heating and cooling (heat pump) is gaining popularity as a heat source for buildings. Schools, homes, and offices in some areas of the United States economically and cleanly transfer heated liquid from underground tubes to/from interior spaces with exceptional efficiency.

d. Solid Waste
(Renewable energy resource)

The heat generated by the burning of solid municipal waste can be used to heat water to operate steam-powered generators. This method also provides for the disposal and recycling of solid waste. Most of the combustible materials are burned in the power plant while the remaining recoverable minerals are recycled (Figure 3-B-9). In Nashville, Tennessee, municipal buildings are heated by underground pipes transferring heat from waste incinerators. This facility has been in operation for 20 years.

In addition to using waste as a source of energy, this system offers a partial solution to the disposal of discarded materials. Solid waste use as a fuel is underused. Therefore, much energy potential for the future involves this fuel. After all, the U.S. leads the world in garbage and trash generated per capita per year.

e. Wood
(Renewable energy resource)

Some scrap wood is used to power boilers in a few mills but virtually none is consumed for general electric generation. Forests would disappear quickly if the U.S. were dependent on wood to fuel electric generation sites.

Photo: Nashville Therman Transfer Corp.

Figure 3-B-9. Solid waste materials being prepared for use as fuel in a steam-powered generating plant.

3.
OTHER ENERGY FORMS

A variety of underdeveloped, emerging energy technologies are probably an indicator of the developing opportunity for a wider array of energy forms rather than primarily dependence on the three nonrenewable fossil fuels (coal for electric generation, oil for transportation, and natural gas for heating buildings). Several developing energy technologies for the second millennium are:

a. Wind

Use of this resource is practical where consistent air currents are somewhat dependable, perhaps on high ridges, the plains or near the seacoast. New technologies use huge light weight, durable plastic rotors and spirals which spin more rapidly and effectively than the old wind equipment for generating electricity (Figure 3-B-10). Most wind operations are small scale. However, several large scale operations are showing promise.

b. Thermal Conversion

Huge mirrors reflect sunlight concentrating this energy to heating water to produce steam which then turns turbines for generation of electricity.

c. Photovoltaic Conversion

Solar cells can convert sunlight directly into electricity without the use of a turbine and generator. Currently this technology is used for satellites and remote areas (roadsigns, weather stations, etc.).

d. Tidal

The ocean's tidal waves can be used to turn turbines not unlike falling water in hydro-electric generation. This technology is limited to sites along ocean shorelines.

e. Fuel Cells

Fuel cells are products of space age technologies. They efficiently use fossil fuels, methane gas, aerobic digester gas, etc., without combustion or pollutant emissions.

These electrochemical cells work like batteries in reverse, reacting fuel with oxygen in the presence of a catalyst to produce electricity with 2 by-products (water vapor and carbon dioxide). They have been used successfully in the USA and Japan for large quantities of electric power.

f. Fusion

Adding another oxygen atom to water (H_2O) results in heat of fusion. This heat converts water into steam for turning turbines for generation of electricity.

Although more far ranging and currently costly in development compared with other energy forms, the wide availability of water makes this a promising asset.

g. Biogas

Methane gas from landfills, lagoons, etc. can be used as fuel. A major southern utility now captures methane from a city landfill, makes steam and sells it to a textile company.

Photo: Northern States Power Company

Figure 3-B-10. Wind powered generators are only practical where consistent air currents are dependable.

C. Considerations for the Future

As economic and fuel availability conditions change, Americans may be forced to consider other energy sources. Some of the factors that must be considered are discussed as follows:

1. Criteria for Acceptability of Energy Alternatives.
2. Energy Conversion.

1.
CRITERIA FOR ACCEPTABILITY OF ENERGY ALTERNATIVES

Criteria for practicality of use and widespread acceptance of alternative energy forms for the next millennium depends on at least four factors:

- Cost
- Safety
- Reliability over time
- Widespread service

Cost. Cost per kilowatt hour is perhaps the most critical factor for acceptance. Currently, photovoltaic, fusion, and other developing technologies can generate electricity costing from $10 to over $100 per kilowatt hour - far more costly than the typical 8¢–11¢ per kilowatt hour presently being charged in the U.S.

Consumers are unlikely to switch to energy alternatives until costs are more competitive. Costs could become more competitive with innovations similar to those that have occurred with fossil fuels and in a variety of other developing technologies (calculators, computers, and microwave ovens) in recent years.

Safety. Consumers in the U.S. expect high levels of safety for employees at generation and transmission sites as well as points of use in homes, workplaces, and communities. Public opinion has not been forgiving of safety issues among utilities for several decades when compared to injury/death from other situations in society.

Reliability over time. Consumers typically expect reliable supplies of fuels for generation of electricity. Outages are inconvenient, costly in downtime and seem to lead to violent, out of control behaviors in some areas.

Widespread service. Generation of electricity is typically needed for thousands or even millions of consumers, businesses, and other customers. An energy alternative which serves only a few customers will not be promising until its development extends across a wider market.

2.
ENERGY CONVERSION

Historically, fossil fuels only gained wide acceptance around the turn of the century–after research and innovation solved affordability issues. Indeed, successful energy conversion in the U.S. from a national dependency on wood, animal power (labor), and wind (for sailing) to fossil fuels was only possible because criteria for practicability were realized.

As fossil fuel supplies in the U.S. diminish and imports increase we face still another energy crisis. Although many Americans are unaware of or even deny the reality of the eminent energy crisis, similarly some consumers and policy makers do not recognize previous effective U.S. energy conversions from the past.

How Electric Energy Is Transformed, Transmitted & Distributed

As electric energy is generated, it is transformed and transported instantaneously through a network of wires to the consumer (Figure 4-1). A distributor of electricity to end use consumers is one of the following types of utility organizations:

- investor owned organizations,
- rural electric cooperative corporation, or
- public or municipally owned power utility services.

The purpose of this section is to explore the methods of transforming and transporting of electricity from the generating facility to end uses in homes, workplaces, and communities. Through this study material, **readers will be able to identify and describe the function of terms and phrases listed in the "Important Terms and Phrases" box** found on this page.

Wires used to transport electric energy are known as **transmission** and **distribution** lines. To avoid excessive losses during transportation, voltage and amperage are altered by the use of **transformers**. How electric energy is transformed and transported is discussed under the following headings:

A. How Electric Energy is Transmitted.
B. How Transformers Work.
C. How Electric Energy is Distributed and Marketed.
D. Standby Generators

Important Terms and Phrases

Distribution lines
Intrautility system grids
Intrautility system linkages
Outage
Overhead distribution
Primary distribution lines
Secondary distribution lines
Standby generator

Standing plans
Stepdown transformers
Stepup transformers
Substation
Transfer switch
Transformers
Transmission lines
Underground distribution

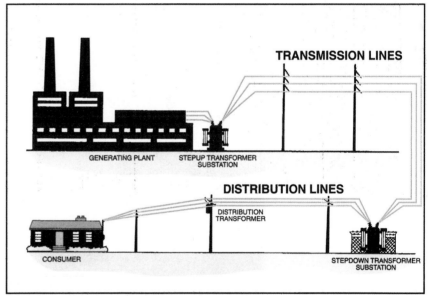

Figure 4-1. Electric energy is generated, transformed, transmitted, and distributed to the consumer.

A. How Electric Energy is Transmitted

Transmission lines. These lines are used to move large quantities of electric energy from **stepup transformers** at the generating plant to the general distribution area. Electric current flows at voltages ranging from 36,000 to 750,000 volts. Typical transmission line voltages are 44,000; 69,000; 72,000; 115,000; 230,000; 325,000; 500,000 and 750,000 volts. Transmission distance may be several hundred miles. These are the lines that are supported usually on large high voltage steel or aluminum towers that extend more than 100 feet in the air (Figure 4-A-1).

As the electric power nears the point of usage, transmission voltage is reduced by **stepdown transformers**. This voltage reduction is done at either a transmission substation, a distribution substation, or both.

Figure 4-A-1. Transmission lines used to transport electric energy from the generating plant to the distribution substations.

B. How Transformers Work

When transporting electric energy, it is necessary to change the relative amounts of voltage and amperage on the line through the use of **transformers**. Basically, a transformer consists of two coils (windings) of wire around an iron core. One coil has more turns than the other. A transformer can be made to "stepup" or "stepdown" the voltage by the way it is connected. When the voltage is stepped up, the amperage is reduced in the same ratio. When the voltage is stepped down, the amperage is increased. In either case (according to the formula and disregarding the transformer efficiency loss, watts = volts x amperes), the amount of power (watts) remain the same (power in = power out). A group of transformers is called a **substation** (Figure 4-B-1).

Transformers are explained under the following headings:

1. How Stepup Transformers Work.
2. How Stepdown Transformers Work.

A substation may contain a group of transformers along with other equipment to manage distribution of electric energy.

Figure 4-B-1. Sub-stations contain groups of transformers.

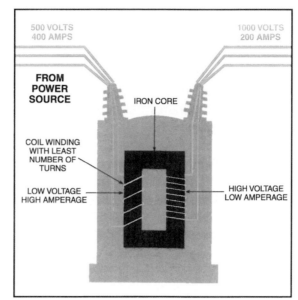

Figure 4-B-2. Principles of a stepup transformer.

1.
HOW STEPUP TRANSFORMERS WORK

Large amounts of electric energy (thousands of volts and amperes) are generated at electric power generation plants. Generators typically generate electricity at over 15,000 volts. Because of the resistance to the flow of electric current through wires (conductors), some of this power is lost while it is being transported. It has been determined that by reducing the amperage, the power loss in transmission and distribution lines is also reduced.

Stepup transformers. These transformers are used at the power plant to increase the voltage and decrease the amperage (Figure 4-B-2). As a result, efficiency is increased and smaller transmission wires can be used.

In the step-up transformer, the power source is connected to the coil with the least number of turns. Electric energy is induced in the second coil. Power induced into the second coil has the amperage decreased in proportion to the increase in voltage. This change in voltage and amperage is proportionate to the number of turns in each coil.

2.
HOW STEPDOWN TRANSFORMERS WORK

Before electric power can be used, the voltage is stepped down and the amperage is stepped up. This is done first when transferring power from distribution lines to service lines. A **stepdown transformer** is used for this purpose (Figure 4-B-3).

In stepdown transformers, the power source (transmission line) is connected to the coil with the greatest number of turns. Power taken from a coil with the fewer turns has decreased voltage and increased amperage.

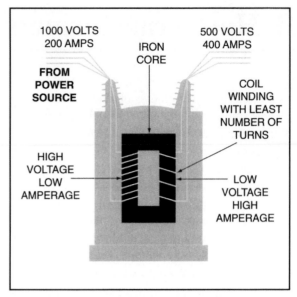

Figure 4-B-3. Principles of a stepdown transformer.

C. How Electric Energy is Distributed and Marketed

The distribution of electric energy includes transporting it from the transmission substation step-down transformer to the customer's electric meter. Distribution is discussed under the following headings:

1. Use of Distribution Lines.
2. Continuity of Electric Service.
3. Marketing Electric Energy.

1. USE OF DISTRIBUTION LINES

Electric energy is distributed by lines that fan out across cities and over the countryside from stepdown distribution substations to deliver power to users. Generally, the two types of distribution lines are as follows:

• **Primary distribution lines** distribute electric energy with voltages ranging from 2,300 to 34,000 volts. Typical primary line voltages are 4160/2400; 12,470/7,200; 13,800/7,980; and 24,940/14,400 volts (Figure 4-C-1).

Distribution transformers step the voltage down for each individual customer from the distribution voltage to the secondary distribution voltage.

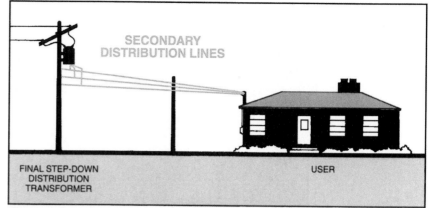

Figure 4-C-2. Secondary distribution lines conduct electric energy from the final stepdown transformer to the user. Final (secondary) voltages are those commonly used, such as 120, 240, 480, and/or 575 volts.

• **Secondary distribution lines** branch off from the distribution transformers and carry electricity with voltages of 120, 240, 480, and/or 575 volts (Figure 4-C-2).

Distribution lines may be constructed by the following methods:

    a. Overhead Distribution
    b. Underground Distribution

a. Overhead Distribution

The traditional method of distributing electric energy to the user employs wires attached to poles above the ground. This

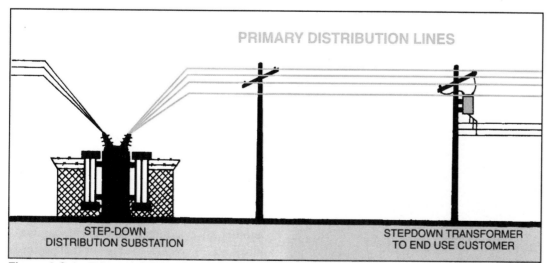

Figure 4-C-1. Primary distribution lines carry electric energy from a stepdown transformer. They usually do not extend more than 25 miles from the substation. Line voltages are usually no higher than 35,000 volts.

Figure 4-C-3. Typical overhead distribution of electric energy.

method is referred to as **overhead distribution** (Figure 4-C-3).

b. Underground Distribution

Many electric distribution lines are now buried below the ground as was service from the first utility company (developed by Thomas Edison in the late 19th century to serve 70 customers in New York). This method is called **underground distribution** (Figure 4-C-4). The primary lines enter the property underground. They are connected to a trans-

Figure 4-C-4. A pad mounted transformer for underground distribution.

former placed on a concrete pad, either above or below ground level (Figure 4-C-4). The secondary lines, also underground, connect from the transformer to the customer's meter. Although more expensive to construct, underground systems result in a neater appearance and are better protected from weather hazards and vandalism.

2 .
CONTINUITY OF ELECTRIC SERVICE

When the flow of electric current through transmission or distribution lines is interrupted an **outage** occurs. Reasons for outages are discussed in other sections of this book. Today, most utility companies have networked to develop (1) **standing plans**, (2) **intrautility system linkages**, and (3) **interutility system grids** to prevent long term outages and their impending consequences.

• **Standing plans** are organized by local utilities in order to manage power distribution during severe widespread shortages. In case of damage to the system due to weather or man-made causes, the utility can prioritize one or more of these standing plans:

- request that consumer voluntarily reduce usage,
- automatically reduce system voltage,
- exercise interruptible power distribution contracts, or
- activate proceedings to rotate blackouts of power.

• **Intrautility system linkages.** In case of damage to a substation, power companies have the ability to link to the next closest substation with reduced interruption in service, sometimes as small as seconds.

• **Interutility system grids.** In the event of a period of near or peak demand, power companies may purchase power from another utility company to prevent massive blackouts. Typically the cost of this power is passed on to the consumer as a cost of doing business.

3. MARKETING ELECTRIC ENERGY

It is the responsibility of the power supplier to deliver electric energy to the consumer. It may be delivered wholesale to another utility (Figure 4-C-5) or retail (Figure 4-C-6). The consumer can then use electric energy according to individual choice. Where population is quite scarce, such as in deserts, mountains, frigid lands, and jungles, the cost of delivering electric power through conventional methods is prohibitively expensive. These expenses can far exceed the revenue generated by the user for years. In isolated terrain, small generators fueled by gasoline or solar energy can supply power for a number of electric products.

Other factors involved in the marketing of electric energy include fuel cost, regulatory agencies and fuel cost adjustments. These are discussed elsewhere in this publication.

Figure 4-C-6. Retail consumers of electricity have power delivered to their homes or businesses through a meter located at the point of use.

Figure 4-C-5. The distribution of electric energy to other power suppliers, municipalities and industrial plants is metered at a substation.

D. Standby Generators

Photo: Jackson Co. REMC

Figure 4-D-1. A transfer switch can prevent death and injury by preventing the flow of electricity onto the utility's lines while the generator is in operation.

Standby generators, or emergency generators are important to rural areas, businesses and services where an interruption of service can be critical. For example, a poultry or dairy operation that relies on electric equipment to power feeding and production activities can be severely affected if the equipment is not supplied electricity.

Depending on its capacity, a standby generator can produce enough power to maintain essential operations and keep refrigeration and heating equipment powered until normal service is restored. Standby generators may be permanently installed in a fixed location or can be a smaller portable unit.

It is extremely important that this equipment be properly connected to existing electric wiring. There are **National Electric Code**® and local electrical standards that must be followed to insure the safety of people and property.

A **transfer switch** must be used when a generator is connected to existing electrical wiring (Figure 4-D-1). This switch is a special switch that prevents the flow of electricity onto the utility's lines while the generator is in use (called "backfeeding"). This switch will also prevent the flow of electricity into the generator when the utility restores service. A transfer switch may be operated automatically or manually.

Determining The Amount And Cost Of Electric Energy Used

Charges for electrical services are usually billed on a monthly basis after the previous month's energy use has been metered (not prepaid) (Figure 5-1). Consumers determine the amount of electricity they use and their resultant charges. Users of electricity choose (1) the number, size, and design of the buildings and the appliances they own; (2) how long equipment and appliances are on, (3) how high or low thermostats are set, and (4) when and how long building doors and windows are open. U.S. consumers have the flexibility to choose between large appliances, homes, and other buildings, or smaller more energy efficient living spaces and appliances. Consumers choose to use energy responsible features and appliances, or to be energy wasteful.

This section provides information about calculating the amount and cost of electricity used. Upon successful completion of this section the reader **will be able to determine the amount of electricity used, calculated by three different methods. The reader will also be able to determine the cost of electrical energy using three different electric rate schedules.**

The reader **will also be able to define and explain the significance of the "Important Terms and Phrases"** box found on this page.

Determining amount and cost of electrical energy are discussed under the following headings:
  A. Determining Amount of Electrical
     Energy Used.
  B. Determining Cost of Electrical
     Energy Used.

Important Terms and Phrases

Billing demand
Capacitors
Check meter disk
  revolutions method
Check meter method
Commercial rate
Cyclometer-type register
Demand charge
Farm service rate
Fuel adjustment charge
Industrial rate
Inverted rate
Kilovolt-ampere reactive
Lagging
Lifeline rate
Meter constant
Off peak rate

Peak rate
Pointer-type register
Poor power factor
Ratchet charges
Rate schedules
Reactive demand
Residential rate
Retail rates
Scrubber technology
Seasonal rate
Special charges
Special rates
Surcharge
Watts and time estimate
  method
Wholesale rates

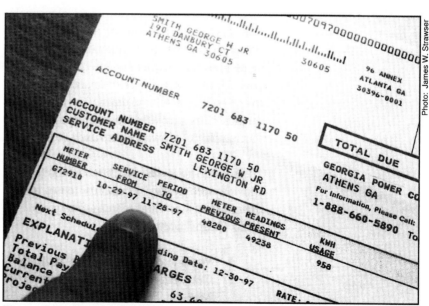

Figure 5-1. Charges for the use of electricity are generally billed to consumers according to meter number. Note meter readings and resulting kwh usage by the consumer over the previous 30-day period.

A. Determining the Amount of Electric Energy Used

Electric bills are usually paid by the month. So in calculating the cost of operation of equipment, the customer needs to know the total number of kilowatthours used during a month's time.

There are three methods for arriving at the amount of electric energy likely to be used by a piece of equipment. They are as follows:

• Watts and time estimate method.
This is especially helpful when an estimate of consumption is needed before equipment is installed. Use the nameplate data on the equipment and estimate the operating time.

• Counting meter disk revolutions.
This is a method of making a quick, accurate check of equipment already in use. Use a regular kilowatthour meter. It has the advantage of measuring accurately the watthours used by equipment where power demands vary from time to time.

• Check meter method.
A check meter is a regular kilowatthour meter installed especially to check electrical usage of one particular piece of equipment. Power suppliers or equipment dealers sometimes supply them where there is considerable question about operating cost over a period of several days or several months. This provides an exact measure of the number of kilowatthours used.

Discussion under the next three headings provide step by step procedures to use for determining how much electric energy is used. Those headings are:

1. Watts and Time Estimate Method.
2. Counting Meter Disk Revolutions Method.
3. The Check Meter Method.

Figure 5-A-1. The nameplate on the equipment for determining power use in watts of electricity (and thus operation cost per hour of use).

1. WATTS AND TIME ESTIMATE METHOD

To determine how much electric energy is used by the watts-and-time-estimate method, proceed as follows:

1. *Determine watts from nameplate. (Figure 5-A-1) (refer to the information found on pages 17-22).*

2. *Estimate number of hours per month equipment will operate.*

3. *Multiply watts by hours of operation to determine watthours (Figure 5-A-2).*

4. *Divide watthours by 1,000 to find kilowatthours of electrical energy used per month (Figure 5-A-2).*

5. *Multiply kwh by local electric rate to estimate cost of operation per month. (The electric utility bills consumers a specific charge per kwh of electrical usage. In the U.S., this rate averages 8¢ or more.)*

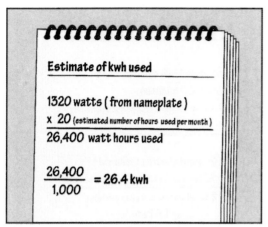

Figure 5-A-2. Estimating the kilowatthours using the watts and time estimate method.

2.
COUNTING METER DISK REVOLUTIONS METHOD

Some meters have flat aluminum disks with black marks along their edges (Figure 5-A-3). This disk turns when electric power is used. The more energy (watthours) used, the faster the disk rotates. A meter can also have a **meter constant**. The constant is shown on the meter nameplate (Figure 5-A-3). A constant of "kh = 7.2" means that for each revolution of the disk, 7.2 watthours have been used (constants will vary with different meters).

Figure 5-A-3. The disk on this meter has a black mark along the edge that can be seen as the disk turns. The "7.2kh" on this meter indicates that with each disk revolution, 7.2 watthours of electricity are used.

To determine how much electric energy is used by counting meter disk revolutions, proceed as follows:

1. *Find the circuit that supplies the appliance to be checked* (Figure 5-A-4). For purposes of this discussion, the appliance we wish to check is a portable oven.

Plug the portable oven into the outlet where the appliance is to be checked, then one after another, trip the circuit breakers or remove the fuses in the panel board until the oven goes off. This breaker or fuse serves the circuit to be checked. Return the circuit breaker to the "on" position or replace the fuse in the fuse box.

Trip all other circuit breakers off, or remove all other fuses, so that all other circuits are off from the individual meter.

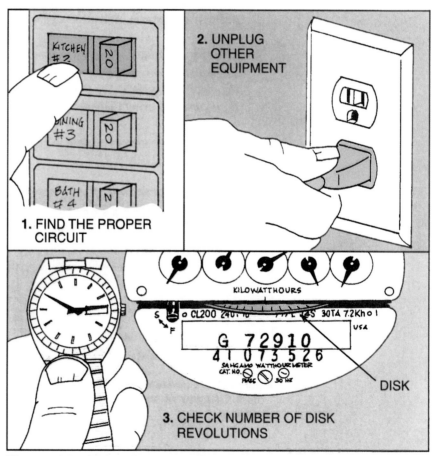

Figure 5-A-4. When using a meter for checking consumption of an appliance, (1) find the proper circuit you wish to check, (2) disconnect all other equipment or make certain all switches are in the "off" position and (3) check the number of disk revolutions over a timed period.

2. *Disconnect any other equipment on the same circuit with the appliance to be checked, or make sure that all their switches are in the "off" position (Figure 5-A-4).*

In Figure 5-A-4, the oven is to be checked. All other appliances which are on the same circuit have been disconnected.

3. *With a watch, check the number of meter disk revolutions over a time period (Figure 5-A-4).*

You may use any period of time. Six minutes makes a good period for equipment that operates continuously. Six minutes is $1/10$ hour and can be easily calculated.

Equipment, such as a refrigerator, that is off part of each hour, should be checked through at least one on-and-off cycle.

4. *Determine watthours of energy used in the selected time period by multiplying the meter constant by the disk revolutions for that period (Figure 5-A-5).*

We observe the meter constant in Figure 5-A-4 to be 7.2 (kh = 7.2).

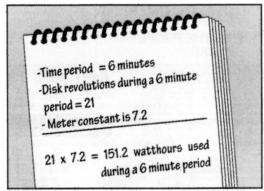

Figure 5-A-5. Computing watthours used from the counting meter disk revolutions method.

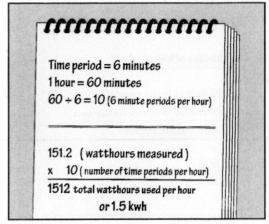

Figure 5-A-6. Computing watthours per hour.

5. *Divide the minutes in the timed period into 60 minutes to determine how many such periods are in an hour (Figure 5-A-6).*

The six-minute period used will equal 10 periods in an hour ($1/10$ hour). If the check period is 15 minutes, there will be four periods in an hour ($1/4$ hour).

6. *Multiply the number of watthours measured in step 4 by the number of time periods per hour. This will give the total number of watthours of energy used per hour (Figure 5-A-6).*

7. *Determine the number of hours equipment is used per month.*

For equipment that is connected to the circuit continuously, calculate the total hours per month (24 hrs. X 30 days = 720 hrs. per month). For equipment that is used only part of the time, such as a microwave oven, toaster or iron, estimate the number of hours used per month. For this problem, assume that the microwave oven operates for an average of one hour each day.

8. *Multiply hours of operation per month by watthours used per hour to determine the number of watthours used per month (Figure 5-A-7).*

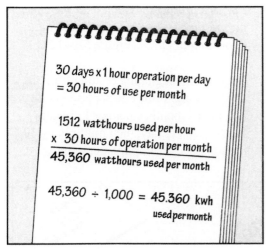

Figure 5-A-7. Computing kilowatthours used per month.

9. *Divide watthours used per month by 1,000 to find number of kilowatthours used per month* (Figure 5-A-7).

3.
THE CHECK METER METHOD

A meter may be installed temporarily to check on the electrical energy used by a piece of equipment. After a check meter is installed (Figure 5-A-8), it is a simple procedure to determine the number of kilowatthours used.

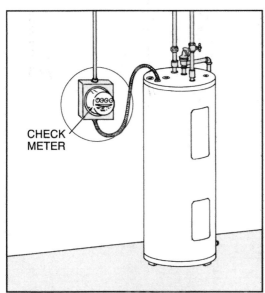

CHECK METER

Figure 5-A-8. Use of a check meter.

Here are the steps:

1. *Note and record the meter reading at the beginning of the check period.*

There are two types of meters. The first type is the **pointer type register** (Figure 5-A-9). Always read the right dial first and proceed left from this dial to read subsequent dials. Alternate dials turn in counter clockwise directions. Reading this type of meter requires skill since the pointer or any one dial rotates in the opposite direction to the one next to it (Figure 5-A-9). You must note the direction in which each pointer moves before recording the number. *Never record a number approached by the pointer. This results in over-reading the energy use.* Instead, read and record the number the pointer has immediately passed.

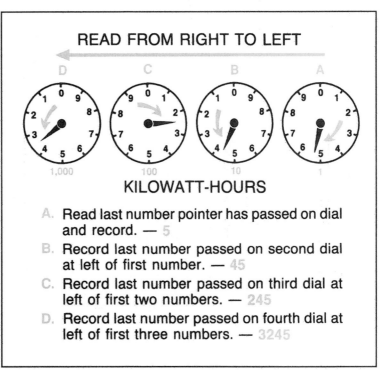

READ FROM RIGHT TO LEFT

KILOWATT-HOURS

A. Read last number pointer has passed on dial and record. — 5
B. Record last number passed on second dial at left of first number. — 45
C. Record last number passed on third dial at left of first two numbers. — 245
D. Record last number passed on fourth dial at left of first three numbers. — 3245

Figure 5-A-9. Proper method of reading pointer type register at the beginning of the check period. Always start with the right hand dial and follow the steps shown.

Figure 5-A-10. A cyclometer type register is read the same way as an automobile odometer.

The second type meter is the **cyclometer type register** which is read the same way as a car odometer (Figure 5-A-10). This type is commonly used by power distributors whose users read their own meters.

2. *Note and record the meter reading at the end of the check period* (Figure 5-A-11).

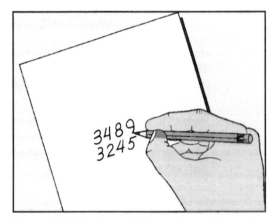

Figure 5-A-11. Record the meter reading at the end of the check period.

3. *For this problem, assume that the check period time is 20 days. Calculate total kilowatthours used* (Figure 5-A-12).

Subtract the beginning reading from the final reading. If the check period is for less than a month, calculate kilowatthours used per day, then multiply by 30 days to get consumption per month.

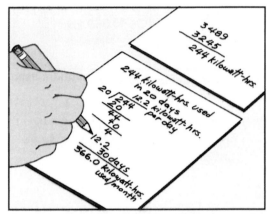

Figure 5-A-12. Compute kilowatthours used from information obtained from meter readings.

B. Determining Cost of the Electrical Energy Used

Each power supplier of electricity has **rate schedules** which show the cost per kilowatt hour unit to users of electricity. The unit cost (of a kilowatt hour) varies across residential, commercial, industrial, and other user categories because of the level of energy demand at a specific time and related equipment and service needed. Rates of investor-owned companies are approved by Public Utility Commissions or their counterparts in each state. Rates of electric cooperatives and public/municipal power districts are approved by their board of directors or city commissions. Many cooperatives and municipals are also governed by commissions.

Electric rates are raised because of increased fuel, power generation, transmission and other costs of doing business. Utilities also lower rates when costs go down. Rates vary from around 6.0¢ per kwh to 18¢ per kwh in the U.S. to twice that in many other countries. Because families pay high electric rates in other countries, they use much less electricity than in our country. Affordable electric rates in the U.S. supports a high quality of life but have unfortunately tended to lead to: (1) wasted energy use, (2) lack of energy responsibility by users, and (3) higher monthly electric bills as a consequence of careless energy use. Because of these problems, one distribution utility has initiated a program emphasizing "use what you need, but need what you use."

Residential, commercial, and industrial customers are charged their respective rate category for a billing period, normally a month. However, the month long billing period will usually not be the same as the calendar month. A meter is read on approximately the same day of each month (with allowances for weekends and holidays) in order to calculate the kwh use which is then multiplied by the respective appropriate rate.

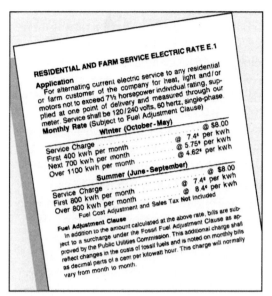

Figure 5-B-1. Summer rates may differ from winter rates and be assessed through seasonal surcharges or special rates.

Special charges may be added in addition to the basic cost of electricity. These may include a charge for minimum monthly service, peak demand charges, fuel cost adjustments, taxes, surcharges, outdoor lights, and burning of fuels with reduced emissions to meet air quality standards. These standards are mandated and enforced by the U.S. Environmental Protection Agency (EPA) at electricity generation sites. Rates charged in summer months may differ from those charged during the winter, either by surcharges or by special rates (Figure 5-B-1). In warm climates, the charges may be higher in summer. In cold climates, the charges may be higher in winter. Higher seasonal rates result from heavy consumer demand reaching peak generation capacity at the electric generation system. If more generation capacity were added, construction and maintenance/operating costs would drive rates above special, peak rates even more.

In general, there are five classifications of rates: **residential**, **farm**, **commercial**, **industrial** and **special**. A specific electric utility may include farm rates with residential or industrial rates, or have a rate of their own. In this publication, farm rates are included with residential rates. Some utilities have the same rates for commercial and industrial customers. Special rates can apply to any of the rates. Some utilities charge higher rates during peak energy use periods such as 7–9 a.m. and 5–8 p.m. (Figure 5-B-2). The higher **peak rates** encourage consumers to shift usage, when possible, to "off-peak" hours when consumer demand for power is lower and the cost to provide service is lower. Further, some utilities organize special programs to encourage consumers to shift some clothes washing, dish washing, bathing, and other power uses to after peak times. For example, a "Wait til Eight" program urges consumers to defer some energy use until after the 5–8 p.m. peak period of demand.

The **residential or farm service rate** is used for lighting, heating, and small motors in a home, on a farm or in a small shop.

Commercial rates are used for professional offices, retailing establishments and restaurants, and institutional facilities (including hotels, hospitals, schools, parks, and sometimes churches). Industrial rates apply to large processing facilities and manufacturing operations.

Electric energy costs are discussed under the following headings:
1. Determining Cost at Residential or Farm Service Rates.
2. Commercial and Industrial Rates.
3. Determining Cost at Special Rates.

1.
DETERMINING COST OF ELECTRICITY USE AT RESIDENTIAL OR FARM SERVICE RATES

Although the discussion will concentrate on the most common rate, the reader should be aware of additional special rates some utilities offer their customers. Most customers will be billed based on the typical residential or farm service rate. However, there are other rates that may apply in particular situations. The local utility should be consulted for information about the availability of any special rate structures and the requirements for qualification to be placed under those rates. Some of these rates are:

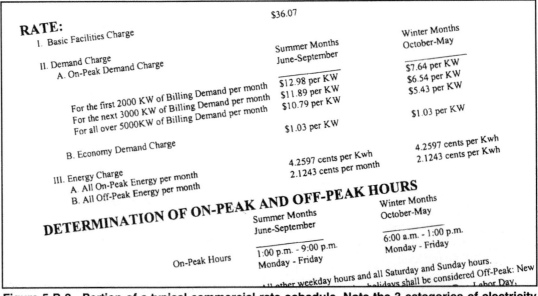

Figure 5-B-2. Portion of a typical commercial rate schedule. Note the 3 categories of electricity charges billable monthly to the end user. See discussion about commercial/industrial rates, page 72.

• **Residential or farm service rate.** A typical residential or farm service rate schedule is shown in Figure 5-B-1. Rates vary widely across the country and from utility to utility, but are usually stated in steps or blocks of energy. These rates provide for a lower cost per kilowatthour after several hundred kwh of electrical energy are used. In order to encourage wiser, more energy conscious decisions, some utilities offer a flat rate with no rate benefit for heavy or wasteful use of electricity.

• **Inverted rate.** In a contrasting method, many utilities now provide a low cost first block to provide low electric rates for those users who conserve electricity (under 1,000 kWh per month). This inverted rate encourages good energy management practices which benefit families and the environment.

• **Life line rates.** Low income families, handicapped individuals and senior citizens may apply for special life line rates. Life line rates are less expensive to the consumer but are not widely available.

Determining cost at residential or farm service rates is discussed as follows:
   a. Understanding the Rate Schedule.
   b. Fuel Adjustment Clause.
   c. Computing Cost of Energy for New Electrical Equipment.

a. Understanding the Rate Schedule

To demonstrate the schedule in Figure 5-B-1, suppose a customer is using 400 kwh of energy each month during the winter. This demand of kilowatthours is in the first block only. Each kilowatthour will cost 10¢ (Figure 5-B-3). This would provide $40.00 in revenue but, when added to the $8.00 service charge, would produce a bill of $48.00.

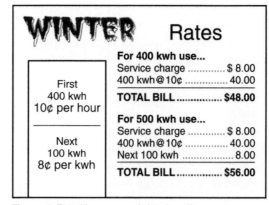

Figure 5-B-3. How you might visualize a rate schedule using 400 kwh and 500 kwh per month.

Suppose that 500 kwh are used each winter month (Figure 5-B-3). The first 400 kilowatt hours would cost $40.00 (400 x 10¢), the next 100 hours would cost $8.00 (100 x 8¢). Adding the $8.00 service charge means the total bill would be $56.00.

A comparison of costs for 1400 kwh used in the summer and winter based on the schedule in 5-B-1, is shown in Figure 5-B-4. During the winter, the first 400 kh would cost 10¢ each or $40.00. The next 700 kwh would cost 8¢ each or $56.00. The last 300 kwh would cost 6¢ each or $18.00. When added together the charge is $114.00. The addition of the $8.00 service charge raises the total winter bill to $122.00.

Figure 5-B-4. How you might visualize the billing difference between winter and summer rates for a customer using 1400 kwh per month.

The summer rate shows that the first 800 kwh would cost 9¢ each, totalling $72.00. The next 600 kwh costing 11¢ each would add $66.00. The charges for 1400 kwh is $138.00. Again, the $8.00 service charge must be added making the total summer bill $146.00.

In this example, the difference in the summer and winter rates is partly to allow for the extra demand for energy caused by the use of air conditioners.

b. Fuel Adjustment Clause

Many electric power customers have noticed a **fuel adjustment charge** added to the electricity bill. This means that the power supplier used coal, oil or natural gas as the major source of energy to generate electricity (to operate the generators). Years ago, these fuels were bought from the less expensive sources. Unfortunately, the least expensive coal and oil contain sulfur and other impurities. These impurities remain after the fossil fuel has been burned and converted to heat. These remaining materials are largely gases which are discharged into the air, thus polluting the air. The Clean Air Act of 1970, passed by the U.S. Congress, requires power suppliers to burn cleaner fuels. Although these fuels contain fewer impurities, they are more expensive to purchase. This means that fuel used by the power suppliers since the law was passed costs more than before. The extra cost of fuel is passed on to the customer through the fuel clause among some utilities.

Generation costs for electricity have also been driven upwards due to another factor. To prevent fuels which are not clean burning from emitting unwanted pollutants into the air, **scrubber technology** is now required by the Environmental Protection Agency (EPA) in order to assure air quality standards for our environment. The scrubbers are placed in the smokestacks at the generation sites to filter emissions into the air. The millions of dollars of cost for this cleaning are passed on to consumers in their energy rates.

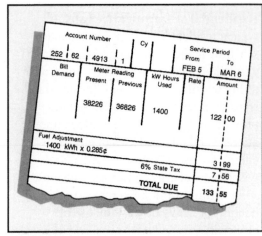

Figure 5-B-5. An example of a monthly electric bill. Note the extra charge for the fuel adjustment.

Additional information about this technology can be found on page 79.

In 1973-74 there was a dramatic increase in the cost of imported oil coupled with the deregulation of government price controls. This resulted in increased costs of oil and natural gas produced in the United States. Costs for coal also rose dramatically in a short period. These rising fuel costs increased the costs of generating electricity, especially electricity then generated with oil and natural gas. The fuel adjustment charge is based on the contracted costs of fuel. The fuel adjustment charge added to monthly bills is a calculation on all kilowatthours used by a consumer each month.

A bill for 1400 kwh in a month with a fuel adjustment of .285¢ per kwh, and the tax added, can then be calculated as shown in Figure 5-B-5.

c. Computing Cost of Energy for New Electrical Equipment

One reason for learning about electric rates is to help in calculating the cost of operating a new piece of equipment. Remember, one way to estimate the amount of energy needed is the watt and time estimate method. Cost of this additional energy for the new equipment can also be estimated as follows:

Calculation 1.

(Based on Summer Rates, Figure 5-B-1.)

1. *Secure the appropriate rate schedule from the local power supplier.*

   Make sure this is the rate which will be used with the new equipment that is to be added.

2. *Check several recent monthly bills to determine about how many kilowatt-hours are being used each month.*

3. *Estimate how many kilowatthours of electricity will be used by the new equipment. Use the following as an example:*

   The proposed new equipment is an attic fan powered by a $^1/_3$ hp electric motor. A $^1/_3$ hp motor requires about 400 watts. Estimate the fan will operate for an average of about 6 hours per day during the hottest summer months. Watt-hours each month can now be estimated.

   400 watts x 6 hours (per day) x 30 days = 72,000 watthours or 72 kwh per month during the *summer months*. (See Figure 5-B-1, Summer Rates, page 67.)

4. *Add these 72 kwh to those already being used.*

   Assume you are now using 400 kwh per month. This will make a total of 472 kwh. The cost can be figured as follows: 472 kwh @ 9.0¢ = $ 42.48.

   Because all of the energy is in the same rate block (Figure 5-B-1), the cost of energy to run the fan is simply 9.0¢ x 72 kwh = $6.48 per month.

5. *To this total ($6.48) add the fuel adjustment cost.*

   Assume a charge of .285¢ per kwh.
   72 kwh x .285¢/kwh = 20.52¢ or 21¢.

Determined cost	=	$ 6.48
Fuel Adjustment charge	+	.21
Subtotal	=	$ 6.69
4% tax	+	.27
Total cost of fan operation	=	**$ 6.96**

per month at 9¢ per kwh.

Calculation 2.

(Based on Summer Rates, Figure 5-B-1.)

1. *Find total cost. (See Figure 5-B-1, Summer Rates, page 67.)*

   Assume when the monthly usage is checked, it is found that 1000 kwh per month were already being used. The total used would be 1000 + 72 (fan) for a total of 1072 kwh per month.

   Cost of power for the added fan would then be 72 kwh @ 11¢ = $7.92.
   Cost can be calculated as follows:

First 800 kwh @ 9.0¢	=	$ 72.00
Next 272 kwh @ 11.0¢		
per kwh (summer)	=	29.92
Total calculated cost	=	**$ 101.92**

2. *Subtract the cost of energy **without the fan.***

Total for 1272 kwh	= $101.92
(as calculated in step 1)	
Subtract cost of fan operation,	
72 kwh @ 11¢	- 7.92
Adjusted total	**= $ 94.00**

3. *To this total ($7.92) add a fuel adjustment charge.*
   Assume a charge of .285¢

   72 kwh x .285¢ = 20.52¢ or 21¢

Cost of fan operation	=	$ 7.92
Fuel adjustment charge	+	.21
Subtotal	=	8.13
4% tax	+	.33
Total cost	=	**$ 8.46**

   This is the estimated total cost per month of using the fan for each month if it is operated at 11¢ per kwh.

2.
DETERMINING COST AT COMMERCIAL OR INDUSTRIAL RATES

It is not within the scope of this publication to provide extensive information concerning commercial rates and the involved and sometimes complicated industrial rates. However, a certain amount of information should be a part of the reader's understanding

Many people have the impression that a commercial or industrial rate provides a lower cost per kilowatt hour than a residential or farm service rate. A better understanding of the factors involved in the determination and application of these rates can be found in the following discussion.

- **Commercial Rates.** These rates apply to small businesses who use up to about 250 kw capacity. Commercial rates usually apply to firms and businesses when there is no residence associated with the facility. Commercial customers usually pay substantially more per kilowatt-hour than do residential users. Some electric utilities place farms (without a residence) on the commercial rate. The cost of this electricity may be 15¢ per kwh or higher. Additional charges can be made for short term usage of large quantities of electricity where several motors are involved.

Commercial rates typically apply to restaurants, service stations, food stores, retail stores, banks, automotive stores, churches, chain stores, and individual shops (Figure 5-B-6). The service may be either single-phase 120/240 volts, three-phase 120/208 volts (most common) or 120/240 volts.

- **Industrial Rates.** These rates apply to very large users of electricity starting at about 250 kw capacity and going up to many thousands of kw capacity. Users are typically manufacturing plants who employ hundreds of employees. The electricity may be delivered by the utility to a substation at the plant at medium transmission voltage (Figure 5-B-7). Some common delivery voltages to industrial customers are 4,160; 12,470/7,200; 13,800 and 24,940 volts. Where the electric utility provides the stepped down voltage, it typically is 480 volts, but can be 575 volts.

Figure 5-B-6. Commercial users include restaurants, small businesses and individual shops.

Figure 5-B-7. Industrial users are large manufacturing plants employing many people and using large amounts of electricity. Many industrial customers provide their own stepdown transformers.

Many industrial users provide the stepdown transformers to supply the voltages needed in the plant. Typically this is 480 or 575 volt three-phase. Since the utility has a very small investment at the plant (requiring minimum maintenance), meter readings for delivering huge quantities of electricity (around the clock, 7 days a week), the kilowatt hour charges will be the lowest of all rates. Some industrial rates will be in the order of 4¢ to 7¢ per kilowatt hour.

The rates may provide penalty clauses for time of day usage beyond a calculated figure for that particular plant. Other charges may apply to plants with numbers of motors operating at little or no load for long periods which will result in inefficient usage of electricity and results in high losses to the utility. A plant that can utilize only 50% of the power supplied requires the utility to provide twice the generation and transmission facilities as one operating equipment at 100% utilization. Kilowatt hour meters do not measure the lower efficiency and, if used by themselves, would only bill for the kilowatt hours being measured. This results in substantial losses to the utility. To offset these losses, additional charges for inefficient use of electricity may be included in the industrial rates. Special metering equipment may also be required. The rate schedule can be involved and complicated and is generally considered beyond the grasp of the readership of this publication.

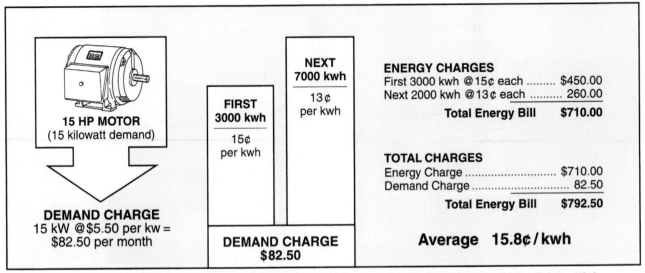

ENERGY CHARGES
First 3000 kwh @15¢ each $450.00
Next 2000 kwh @13¢ each 260.00
           Total Energy Bill $710.00

TOTAL CHARGES
Energy Charge $710.00
Demand Charge 82.50
           Total Energy Bill $792.50

Average   15.8¢ / kwh

15 HP MOTOR
(15 kilowatt demand)

DEMAND CHARGE
15 kW @$5.50 per kw =
$82.50 per month

**FIRST
3000 kwh**
15¢
per kwh

**NEXT
7000 kwh**
13¢
per kwh

**DEMAND CHARGE
$82.50**

Figure 5-B-8. Demand charges are sometimes added for large uses of power for short periods of time. High demands by large users affects voltage, potential for blackouts, and much underused electricity transmission and distribution equipment when the demand is low.

3.
DETERMINING COST AT SPECIAL RATES

There are at least seven common special rates. They are as follows:

    a. Demand Rates.
    b. Poor Power Factor Rates.
    c. Surcharges.
    d. Irrigation Rates.
    e. Special Rates.
    f. Off Peak Rates.
    g. Retail vs. Wholesale Rates.

a. Demand Rates

Electric rates with demand charges recover capital costs while kwh (energy) charges recover operating costs. This rate structure equitably recovers the cost to provide electric service to each customer by their own demands on the system.

Special rates may be applied to equipment which operates only in selected seasons, such as irrigation systems, drainage pumps, grain drying bins, and cotton gins. Also, some utilities use special rates for churches, seasonal resorts, and schools which do not use energy every day of the week when the electric equipment servicing the building stands idle. In these cases, rather large

amounts of power are required for short periods of time. This use may require larger transformers, heavier power lines, additional safety devices and greater generating capacity and reserves. Service and equipment costs to the power supplier may be so great that the revenue from electrical energy at regular rates does not justify the increased investment by the power supplier.

To overcome this loss, the power supplier adds an extra section to the rate. This rate is called a **demand charge** (Figure 5-B-8). It is based on the largest number of kilowatts demanded for a 30-minute period over the past month. It is also called **billing demand**. One method is to charge at least 95 percent of the demand set during the summer months regardless of how little energy may be used during the other months or at least 60 percent of the demand set during winter, whichever is greater. This billing demand is charged for 12 months.

Demand charges are sometimes called **"ratchet charges"** because once the demand is set, this charge cannot be reduced for a year regardless of how little electricity is used. Customers of all types can be subject to demand charges and some try to control peak demand to avoid these charges by using load management principals.

Large installations now use very sophisticated computer-controlled energy management systems to limit their demand. When the demand approaches its peak, various loads are switched off on a priority basis so that a new peak is avoided. For example, using remote control switches a university can turn heating systems off for 5 minutes in each building in a rotating fashion in order to manage its energy demands in near peak status. Some utilities now install switches on home air conditioners, water heaters and pumps to minimize the demand peaks on their system.

A typical demand charge is $5.50 per kw per month. A 15 kw demand would cost 15 x $5.50 per month or $82.50 per month (Figure 5-B-8).

b. Poor Power Factor Rates

The bill may also include additional charges when the customer has a **poor power factor** (less than 85% efficiency of transporting power known as **lagging**). The utility has to set special meters to determine poor power factor which is usually caused by large amounts of motor loads. Special equipment called **capacitors** can be installed to improve a poor power factor.

In the section "How Electric Power is Measured," (page 17) power in AC circuits was defined as volts x amperes x power factor. The power factor is a fraction between 0 and 1.0. The power supplier is paid for energy that consumers use, which is power x time. A low power factor does not reduce energy consumed. The voltage and current must be the same as with a high power factor. Thus, it is to the power supplier's benefit for the power factor to be close to 1.0.

c. Surcharges

Some utilities increase the total bill by a specific percent during the summer or winter months, when demand is the highest. This is

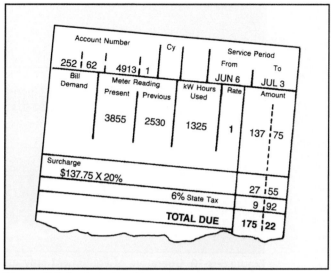

Figure 5-B-9. A method for computing cost when a surcharge is added.

called a **surcharge** and is usually 15 to 25 percent of the bill. This is a convenient method of providing different rates for imbalanced, uneven demand on the utility system during summer and winter. An example of calculating cost of energy supplied under a surcharge is shown in Figure 5-B-9.

d. Irrigation Rates

Special irrigation rates are provided by some utilities for those customers who turn off their irrigation systems during times of peak demands on the electric system. Irrigation pumps may be turned off by remote control switches which may be operated by individuals or the electric utility. A typical method is to turn the pumps off during very hot summer afternoons from noon to 7 p.m. Interestingly, this is the time of day when high heat causes much evaporation leading to irrigation inefficiencies.

An example of an **off-peak rate** would be a service charge of 85¢ per horsepower per month plus 4.9¢ per kwh while other loads would be charged at regular, but higher kwh rate plus $5.00 per horsepower per month. Many times off-peak rates can reduce electricity costs by one-half.

Figure 5-B-10. Operators of equipment only used seasonally want to pay all their costs during this period of time.

e. Special Rates

Large users of electricity such as schools, which use large quantities of electrical energy, may have their own special rate. Based on the individual utility, special consideration may be given to any group or type of users. Some facilities such as cotton gins and peanut mills operate only in the fall to early winter each year (Figure 5-B-10). The ginners want to pay all their costs during the ginning season. This would be a seasonal rate.

f. Off Peak Rates

Some utilities provide a special price break for using electricity at times of low electrical usage on their system called **off peak rates**. The time of greatest electrical usage is typically the hottest time of day in the summer in the south and the coldest time of night or morning in the north. The exact time for this condition will vary from utility to utility. The time of greatest electrical energy usage on a system is called **peak time**. Electric utilities may offer special rates for off peak usage of electricity. This may be based on time of day or season of the year. For example, some utilities charge a different rate for electricity used in the summer months as compared to the winter months (Figure 5-B-2).

g. Retail vs. Wholesale Rates

Retail rates apply to all end users of electricity regardless of their size and their usage of electricity. Regardless of the electrical energy charge rate applying to the electrical service, the customer who actually uses the electricity is a retail user. Therefore, residential, commercial, industrial, farm, and special rates are all retail rates.

Some electric utilities sell electricity to cities or other electric utilities who, in turn, distribute the electricity to the retail customers. The rates used to determine the costs for the electricity are called wholesale rates and are not available to end use customers.

How Environmental Protection Is Provided

As we consumers experience numerous benefits of electricity, we also realize that man must be concerned and careful about protecting the environment we all currently live in. Waste products that come about as the result of generating electricity must be managed wisely as an investment for future generations. The residue from the burning of fossil fuels and disposal of nuclear by-product materials must be carefully planned in order to ensure a desirable quality of life in the near and distant future. Users of energy should make energy credible decisions if behaviors are to lead to positive results. On the other hand, counterproductive energy wasteful behaviors can lead to costly consequences.

The purpose of this section is to familiarize the reader with the challenges faced by the power generating companies when coping with the ever increasing electrical needs of society while protecting the water, air and soil we all come in contact with every day.

As a result of the study of this section **the reader will be able to name the methods used to remove or handle the by-products of electric generating plants. In addition, the reader will be able to identify and give a brief description of the terms and phrases found in the "Important Terms and Phrases" box** found on this page.

Recreational facilities are an integral part of planned electric power generation sites. Lakes and reservoirs at hydroelectric dams provide for swimming, fishing, boating and skiing in addition to flood control. The surrounding areas are developed for camping, hiking, also homes, cabins, lodges,

Important Terms and Phrases

Ash basin
Bottom ash
Cooling lakes
Cooling reservoirs
Cooling towers
Electrostatic precipitators
Excess heat
Exhaust stack

Fly ash
Fuel reprocessing
High-level waste
Low-level waste
Nitrogen oxides
Scrubbers
Spent fuel
Sulfur dioxide

motels, restaurants, grocery/supply stores, and other recreational options. Great efforts are made by electric power suppliers to protect this environment from any harmful by-products of electric power generating plants. The electric power industry spends billions of dollars each year for this purpose. Some of the methods used are discussed under the following headings:

A How Heat is Removed.
B. How Sulfur Dioxide and Ash are Controlled.
C. How Nuclear Waste is Handled.
D. How Hydroelectric Plants Maintain a River's Vitality.

A. How Heat is Removed

Steam powered plants develop **excess heat** or waste heat. This heat must be removed whether it is produced from burning fossil fuels or fission of uranium atoms (nuclear fuel). The steam must be condensed to water so that it can be recycled. The water that is cycled through the condenser to cool the steam must be returned back to its source which may be a:

- **cooling tower** (Figure 6-A-1) or
- **cooling lake, reservoir or river** (Figure 6-A-2).

Once in the cooling tower, or reservoir, natural processes transfer the heat from the water to the atmosphere. This cooling water is returned to its source (lake, reservoir) unchanged except that it is somewhat warmer.

The temperature of the returned water must be within the guidelines established by state and federal agencies to protect aquatic life and is closely monitored by the power supplier. Generally, the water used from a river or reservoir must be returned "cooled" and within $\pm 2°$ of the inlet temperature. In some cases, heated water can be used to heat greenhouses in cooler weather, extending the growing season.

Figure 6-A-1. A generating plant with large cooling towers designed to dissipate waste heat.

Figure 6-A-2. The water taken from a lake, reservoir, or river can be recycled after it has been used to generate steam. The steam is condensed back to water and cooled before it is returned to the source.

B. How Sulphur Dioxide and Ash Are Controlled

Coal burning electric power plants emit **oxides of sulfur** (termed SO_x)* and **oxides of nitrogen** (NO_x)* as well as **fly ash** as part of the combustion process. Appalachian coal tends to have a high sulfur content, whereas Rocky Mountain coal can have less sulfur. The nitrogen oxides form when nitrogen in the air combines at very high temperature with oxygen. The extent of these chemicals in emissions depends partially on the kind of coal burned. Low sulfur coal produces less emission up **exhaust stacks** but may cost more to acquire and is likely to be less available in the future. Sulfur dioxide and nitrogen oxide emissions are regulated to protect public health and the environment through federal government standards enforced by the U.S. Environmental Protection Agency.

Emissions up exhaust stacks may also be controlled by the use of devices known as **scrubbers.** This method of control passes the flue gas through a sprayed water/limestone mixture to remove the majority of the sulfur dioxide before emissions are released up the stack and into the atmosphere (Figure 6-B-1). Stacks are constructed tall enough so that emissions can be diluted in atmosphere high above the land below.

Scrubbers are significant to consumers in a secondary way. The cost of their construction and operation involve millions of dollars which is passed on to consumers in the form of higher utility rates. The cost of scrubbers is a cost of electricity generated from coal.

*The chemical expressions in parenthesis are SO followed by the subscript "x", and NO followed by the subscript "x". The "x" can equal 1, 2, 3, or 4 depending on the condition in the furnace.

Photo: Georgia Power Co.

Figure 6-B-1. So called "scrubbers" are designed to modify the emissions from fossil fueled generating plants before pollutants are released into the atmosphere.

The watery limestone mixture produced by scrubbers is usually stored in large sludge basins (or ponds) that require considerable added land area at the plant site. Some of this sludge can be treated and sold as gypsum to the construction industry. However, most is simply stacked and stored on the land.

Figure 6-B-2. Electrostatic precipitators are used to remove ash particles from coal fired plants before the emissions are discharged into the atmosphere.

Electrostatic precipitators are used by most coal fired power suppliers to remove ash particles before they are exhausted into the atmosphere (Figure 6-B-2). The ash (in the smoke on its way to the exhaust stacks) is subjected to an electrical charge. The charged particles are attracted much like a magnet to plates with an opposite electrical charge. The precipitators remove up to 99% of the solid matter which could be emitted through exhaust stacks. Residue (solid waste) from this process is collected in an **ash basin** or ash pond for further processing and disposal or reuse in manufacturing of other products such as an additive in the making of concrete blocks. Furthermore, the material collected at the bottom of furnaces, called **bottom ash**, is recycled and used as sandblasting material and for granules used in the manufacturing of roofing shingles.

C. How Nuclear Waste is Handled

Almost all radioactive wastes from power plants can be placed into either of two categories:

- low level waste
- high level waste

Low level wastes. This category of nuclear waste includes such things as wiping rags, mops, wet and dry filters, tools, some protective clothing items, and even valves and piping that have been replaced during regular plant maintenance. The level of contamination is low (decaying to background levels in less than 500 years) and permits the wastes to be handled with ordinary (but very clean) methods. These items are typically shipped to a contractor who separates reusable materials and then forwards the remainder for underground burial at specially prepared sites for final disposal. Disposal fees can be based on the volume of material, so the waste is typically compacted before shipment.

Low level waste is produced in all 50 states by more than 12,000 companies and institutions. The utility industry disposed of 2.1 million cubic feet of low-level waste in 1980. In 1993, the volume declined to about 386,000 cubic feet even though the number of nuclear power plants increased by more than 50 percent. Nuclear power plants account for less than 40% of the volume of low-level waste disposed of in the U.S.; industrial sources account for just over half the volume, with medical, government, and academic facilities accounting for the rest.

The National Research Council's requirements place restrictions on the types of waste which can be disposed. Current low level disposal uses shallow land burial sites without concrete vaults.

High level waste. The high level waste from nuclear power plants consists entirely of used nuclear fuel in solid form (Figure 6-C-1). The hard, ceramic-like fuel pellets consist of

Photo: Duke Power Co.

Figure 6-C-1. High-level nuclear waste from nuclear powered plants must be stored in approved containers for movement to long-term storage areas.

only 4% Uranium 235, the material that actually fissions to produce heat. A typical nuclear power plant produces about 20 metric tons of used fuel every year. All of the nation's nuclear power plants combined produce about 2,000 tons of used fuel each year. In more than 35 years of operation, all of the used fuel produced by the nuclear energy industry totals only about 32,000 tons. While that may sound like a lot, those 32,000 tons would cover an area about the size of a football field to a depth of about 12 feet, if the fuel assemblies were stacked side by side and laid end to end.

Today, used fuel is safely stored at the nation's nuclear power plants in either steel-lined, concrete vaults filled with water (used fuel pools), or in above-ground steel or steel-reinforced concrete containers with steel inner canisters (dry casks). On-site storage is an interim measure, however. While the Nuclear Regulatory Commission has determined that used fuel can be stored safely at existing plant sites for 100 years, a single, scientifically engineered repository for all of the nation's nuclear waste is also in the making.

Most scientists agree high level nuclear waste can be safely disposed of in properly selected and constructed underground vaults over 1000 feet below the surface of the earth. Such a site must be isolated from concentrations of people and the rock must be stable and not subject to rising underground water levels.

The plan for handling high-level waste is to put the used fuel assemblies into long, round containers made of steel and concrete. Seamless, failsafe containers are designed with suitably thick walls to control the radioactive wave lengths released by nuclear wastes even during transport. These canisters will be transported to the disposal site and be placed underground in specially prepared holes where they can be monitored and kept safely for thousands of years. At this time a site in Nevada is being studied and considered for this use.

Internationally, a few countries have chosen to break down their used fuel and recycle it (called **fuel reprocessing**). Even when used fuel is put into a storage pool, about 13% of the original uranium (Uranium 235 isotope) content remains unused.

By reprocessing the used fuel, this 13% can be recycled into fresh fuel and put into another reactor called a "breeder reactor". Leftovers from the process, however, include plutonium--the key ingredient in nuclear weapons. Isolating the plutonium in the reprocessing cycle makes it available for atomic weapons use, which is why President Jimmy Carter signed an Executive Order in the 1970s prohibiting any reprocessing in the United States. France and the United Kingdom are still reprocessing fuel. Other nations needing large amounts of affordable electric power for commercial use, such as India and China, have not adopted this policy.

D. How Hydroelectric Plants Maintain a River's Vitality

Hydroelectric power plants typically include a dam to control the flow of water in a river, a powerhouse containing turbines and generators to make electricity, and a large reservoir of water behind the dam. The turbine-generators in the powerhouse are turned on and off as needed to meet electrical demands. The generation process is fairly simple and produces no visible pollutants or non-visible hazards. However, animal species that depend on the river's food chain experience several environmental impacts that can have serious consequences if not dealt with responsibly. The major impacts are:

- loss of dissolved oxygen
- loss of marine habitat and natural migration paths

Dissolved oxygen. Marine creatures get their life from oxygen dissolved into water as it rolls over rocks and down waterfalls. The reservoir behind a dam, however, does not provide these natural ways of getting oxygen. Furthermore, as a hydroelectric plant starts and stops producing power, the water tends to become stagnant. Oxygen at the bottom can be depleted, forcing fish to move toward the warmer surface water to compete for the limited supply. Inevitably, some species suffer. This same dissolved oxygen is also consumed by dead trees and processed sewage introduced all along the river's path.

Today's hydroelectric plant engineers have found ways to reintroduce oxygen into the water both as it passes through the turbines and in the riverbed just downstream of the dam. In some cases, liquid oxygen is actually bubbled into the water before it passes through the turbines. Forced air can also be injected into the water as it enters the turbines or immediately afterward as it rushes out into the riverbed. All of these techniques require equipment that either uses electricity or reduces the amount of electricity that is produced in the powerhouse.

Other plants have constructed long weirs downstream of the dam to produce an artificial waterfall that puts oxygen back into the water. Properly designed, these weirs can also keep the area below a dam filled with water long after the generators are turned off, providing a healthier environment for fish. The construction and maintenance of these weirs is included in the cost of generating electricity.

Marine habitat and migration. Disturbing the natural flow of a river is something like building a superhighway through the middle of a nice housing development--the neighborhood suddenly stops nurturing life and becomes a threat. An example is a flowing stream used by fish to travel upstream every year to spawn their young. They need the rushing water and the access it provides to breeding grounds. But one or more hydroelectric plants block the migration path and eliminate the rushing water. One solution is to provide fish ladders for adult salmon migrating upstream during the spawning season, and bypass channels for juvenile salmon migrating downstream so they can reach the ocean, (Figure 6-D-1).The natural process is not completely stopped. The cost of passing water around the plant without generating electricity then becomes a cost of doing business. The presence of fish in the river and lakes, however, will support recreation industries that provide jobs and enrich the area's economy. This is another example of the trade-off among energy needs, the environment, and our quality of life.

Figure 6-D-1. This fish ladder installation on a river in Washington shows both upstream and downstream fish passage technologies. The fish ladder on the lower level allows adult salmon to migrate upstream, while the bypass above it carries younger fish to a safe point downstream of the turbinies. The loop helps slow the water flow to prevent fish injuries.

Photo: John McKern, U.S. Army Corps Engineers

Glossary

Alternating current. Current (amperes of electricity) that flows in first one direction, then fluctuates to 0 amperes in a circuit. This fluctuation typically occurs 120 times per second (or 60 complete cycles per second) in U.S. electrical products. The current that flows through lights, motor driven equipment and appliances such as refrigerators, heating appliances, electronic devices and equipment in the home.

Ampere (amperage). Electron flow (current) along a conductor. When 6.28 quintillion electrons pass any particular point in a circuit in one second, this is called an ampere or amp. Abbreviated "A".

Ampacity. Word combining ampere and capacity. Expresses in amperes the current carrying capacity of conductors of electricity.

Arcing. Undesirable leaping of electrical current from an outlet to prongs (conductors) of an electrical plug; caused by plugging/unplugging an electrical item when the switch is "on" or if there is no switch to turn the electrical item "off".

Circuit breaker. Protective device installed in the circuit to break the flow of electricity when current over a conductor exceeds a predetermined safe limit A "switch" interrupts current to prevent unsafe levels of electricity over wiring in a building or product.

Circuit. The path followed by the flowing electrons (current) from the point where they leave the electricity generating facility until they return to it in a closed loop or path.

Coal. The most widely used fossil fuel for producing steam used to generate electricity in the U.S..

Conductor. A wire or other material, typically metal, that allows electrons to move readily offering low resistance.

Cooling tower. Tower designed to transfer, through natural processes, the heat from water used to produce steam which powers electric generators. When water is cooled it is returned to its source.

Copper. Typical , recommended conductor of electricity; the third best conducting material (after gold and silver) and better than aluminum and other metals. For safety, wiring must be sized proportionate to the amount of electric current to be conducted to the end product use.

Current (or flow of current). The continuous flow of electrons in a circuit (in a conductor) measured in amperes (unit of electron flow).

Cyclometer-type meter. An electric meter that employs a counting system that is read the same way as an automotive odometer.

Delivery line. Wire bringing negatively charged electrons from the electrical source to the point of use.

A complete circuit must have a delivery wire or conductor and a return wire or conductor.

Demandside management. A program or strategy to encourage consumers to use the energy they need but also need what they use. Consumers can choose efficient and energy responsible equipment, building products and user behavior.

Digital electric meter. A newer type electric meter which prevents typically over-read or under-read electricity use at the consumer's location.

Direct current. Current that flows in only one direction in a circuit. This type of current is used in flashlights, portable radios, cameras, boats and automobiles.

Distribution lines. Electric lines that deliver electricity from a stepdown distribution substation or stepdown distribution transformer to the final stepdown transformer at the customer's premises.

Electric circuit. The controlled flow of electrons over a closed conducting path or loop.

Electric meter. A device used to measure the number of kilowatthours used over a specific recent period of time by a consumer.

Electric hazard. An unsafe electrical condition involving wiring and/or equipment. Hazards pose risks of injury and/or loss of property and human life. Risks pose the question of when, not if injury or loss will occur.

Electric pressure. Pressure that results from an imbalance of electrons in the delivery wire and the return wire in a circuit. See volts and voltage.

Electric outage. Condition caused when the electrical current flow from the utility system ceases to a building.

Electrical fault. Unsafe condition that occurs when there is a break in the insulation of a conductor or appliance permitting the appliance and any attached equipment to become energized.

Electricity. The flow of electrons from one atom to another.

Electron. A very small negatively charged particle which can flow from one atom to another in a conducting material.

Electrostatic precipitator. A device used to remove smoke, soot and metals from emissions from a fossil-burning electric generation plant. These devices use electrically-charged metal plates to capture small particles in much the same way a television screen attracts dust in a house.

Energy. The ability to do work, make objects move or produce heat or light.

Energy guide. Sticker affixed to major appliances that provides the estimated cost of operation of the appliance at a certain cost per kwh over a stated time period.

Exhaust stacks. Structures designed to funnel emissions from fossil-burning electric generation plants into the air

Fault. See "Electrical fault.".

Fissionable fuel. Atom splitting fuel such as uranium.

Fly ash. Material produced by the burning of coal for electric generation.

Fossil fuels. Coal, oil, and natural gas mined or piped from the earth. Nonrenewable energy forms.

Fuel cells. Products of space age technologies. They efficiently use fossil fuels, methane gas, aerobic digester gas, etc. without combustion or pollutant emissions. These electrochemical cells work like batteries in reverse, reacting fuel with oxygen in the presence of a catalyst to produce electricity with 2 byproducts (water vapor and carbon dioxide). They have been used successfully in the USA and Japan for large quantities of electric power. Smaller versions are being tested for homes, farms, business and even automobiles.

Fuse. Safety device that prevents excess current from flowing in a circuit. Fuse sizes must be matched to the size and materials of the conductor they protect.

Fusestat. A fuse type device that has a built-in quality to absorb momentary surges of current, then return to a lower current. For example, a motor will require a "surge" to start, but once started, will return to reduced normal current flow.

Fusetron. A fusestat type device with a ceramic base that screws into a metal adapter designed to limit a replacement fuse to the former fuse size (amp capacity).

Generator. A device which produces electricity from a renewable (falling water, geothermal, wind, solar, etc.) resource or from a non-renewable energy form such as coal, oil, natural gas and uranium.

Generating unit. A complete electric generating system. It includes the fuel supply, a turbine and a generator. There may be several units at a plant.

Geothermal. Pertaining to the heat in the earth.

GFCI. A protective device that is sensitive to faults of current flow. Generally used where there is a greater chance of electrical shock from contact between water and electricity, e.g., in baths and kitchens. Its purpose is to turn a circuit "off" when electric current flows in other than the planned path. It turns off quickly enough to prevent electrocution. Limits stray current to 8 milliamps.

High-level waste. Nuclear waste containing long-lived radioactive isotopes that must be isolated for long periods of time to be rendered harmless.

Horsepower. A unit for measuring the power of motors, equal to 746 watts.

Hydroelectric generators. Generators that are powered or turned by the presence of water flowing from a higher level to a lower level.

Insulator or nonconductor. Any material that will not conduct electricity because it will not release its own electrons or allow other free electrons to be exchanged between atoms.

Inverted rate. A low cost rate or schedule that provides a lower first block charge for those users that consume small amounts of electricity. Used to encourage conservation of energy.

KVA. 1,000 x volts x amps.

Kilo. A metric term meaning 1,000.

Kilowatthour. Unit used for metering and selling electricity. Unit of power (1,000 watts) used in an hour (unit of time). Abbreviated kwh.

Lagging. Condition of power transmission by utility company to customers. If efficiency of transporting power is less than 85% (power factor), then the utility company can install special electric meters and assess reactive demand charges.

Life-line rates. Special electric rates designed to assist handicapped, aged or low-income consumers similar to the inverted rate.

Lightning. Life threatening condition caused by the imbalance of electrons in the atmosphere between positively and negatively charged atoms.

Low-level waste. Nuclear wastes that are generally contained in those chemicals and in and on items that are contaminated with low levels of radioactivity. These wastes contaminate clothing, tools and filters used in the process of generating electric energy with nuclear fuel.

National Electric Code®. Published minimum standards to encourage effective safe electric wiring procedures, types of materials, sizes of components which, if followed, can enhance quality of life, and in particular electrical safety.

Nameplate. A metal or plastic label attached to a motor or appliance providing information such as rating in amperes, volts, watts, hertz, horsepower, manufacturer, model number, serial number, etc.

Nitrogen oxides. Chemical emitted by coal burning electric power plant.

Nonrenewable resources. Fixed sum resources that, once used reduce the remaining available supply for the future. Examples are coal, oil, natural gas and uranium.

Nuclear fission. The splitting of atoms which produces large quantities of heat useful for efficient generation of electricity.

Nuclear fuel. A uranium fuel core used to generate heat in a nuclear plant by fission.

Nuclear reactor. An assembly consisting of fuel assemblies, control rods and a coolant housed in thick concrete housing wherein atom splitting occurs which produces heat useful for converting water to steam which turns generating equipment.

Ohm. A unit of electrical resistance (the energy required to force electrons from one atom to the next) in a conductor.

Overhead distribution. Traditional way of distributing electricity using wires attached to poles above the ground rather than underground.

Parallel circuit. A circuit that provides for dividing the current flow to each light or appliance on the circuit. If one light burns out, current will continue to flow and function in other lights on the circuit.

Peak demand. Time period when the greatest need for electricity by users occurs on the electrical distribution system.

Phase converter. A device powered by single-phase electricity that produces three-phase electricity to operate three-phase electric motors and other electrical equipment.

Pointer-type register. An electric meter that employs clock-type dials used to calculate the amount of electricity used by a consumer during a given time period for billing energy costs to the consumer.

Power factor. A unitless fraction used in conjunction with voltage and amperage to produce a watt. The power factor is dependent upon the type of current (direct or alternating) and the type of device the current is flowing through.

Primary distribution lines. Power lines that distribute electric energy with voltages ranging from 2,300 to 34,000 volts.

Reactive demand. The highest 30-minute kilovolt-ampere reactive (kVAR) during the month resulting in a cost of doing business charge to commercial or industrial customers with a low power factor.

Renewable resources. Energy forms whose supply does not diminish after a hour's, a day's or a year's use. Examples are solar, tides, wind power, geothermal and water power.

Residential rate. A schedule of charges for the amount of electricity used in a month. Designed for homes, small businesses and farms.

Resistance. The energy required to force electrons from one atom to the next in a conductor (wire). Measured in ohms. High resistance to conductivity can result in excess, dangerous heat in wiring.

Scrubber. An air pollution reducing device that sprays wetted lime powder into the hot exhaust to neutralize acid gases from fossil fuel generators.

Secondary water system. Low pressure water system designed to absorb the heat created by a nuclear reactor turning the water into steam that is piped to the generator to produce electricity.

Secondary distribution lines. Power lines that branch off from primary distribution lines carrying electricity with voltages ranging from 120 to 575 volts.

SEER. Seasonal Energy Efficiency Ratio. A sticker affixed to heating and air conditioning equipment indicating a numerical rating showing the relationship between the output of a given piece of equipment to the amount of input energy.

SHPF. Season Heating Performance Factor. A numerical rating for the heating efficiency of electric heat pumps.

Seismic fault. A break in the earth's crust where naturally heated water or steam can be accessed for the generation of electricity or for heating buildings directly.

Semi-conductor. A substance that is neither a good conductor or a good insulator. The free electrons are held rather tightly posing resistance to current flow, producing heat for cooking, heating water, drying hair and clothing, etc.

Series circuit. A circuit in which all of the current flows throughout the entire circuit. If one light in this type of circuit burns out, the circuit is broken and other lights on the circuit will not operate.

Short circuit. Occurs when 2 conductors come in contact with each other without going through an appliance, thereby reducing the work done and increasing current flow to a dangerous level.

Single-phase power. Power supplied to homes and small businesses through one transformer, two delivery wires and one neutral (return) wire.

Solar energy. Energy from the sun. A renewable energy form available in passive, active and photovoltaic technology—from affordable to developing technology respectively.

Special rates. Rate schedules designed to accommodate special uses of consumers. Examples are irrigation systems operated in selected seasons, churches and schools that use large amounts of power for short periods of time in contrast to the utility components which are under-used most of the time.

Spent fuel. Nuclear fuel that has been used to produce the steam in a nuclear power plant.

Standby generator. An emergency generator that can be used to provide electrical service when there is an outage because of damage to the utility's lines or an interruption of service. A transfer switch must be used with a standby generator.

Steam power. Electrical power generated through the use of steam created by heating water with an energy form.

Stepdown transformers. Transformers that are designed to decrease the transmission and distribution voltage and increase the amperage for users.

Stepup transformers. Transformers that are used at the power plant to increase the voltage and decrease the amperage, lessening the power loss in transmission and distribution lines.

Substation. A group of stepup or stepdown transformers.

Sulfur dioxides. Chemicals emitted by coal-burning electric power plants.

Surcharges. Additional charges on electric bills by specified percentages or the results of higher fuel costs.

Surge protector. Device to protect electrical equipment from damage due to power surges or spikes of electricity and lightning.

Three-phase power. Three phase power is commonly used by industry for the operation of larger electric motors. Three-phase service is supplied through two or three transformers and four wires.

Transfer switch. A special switch used in conjunction with the installation of a standby generator. It is designed to prevent the flow of electricity into the utility's lines while the generator is in use.

Transformer. A device used to change voltage and amperage on line, either up or down. See stepup/stepdown transformers.

Transmission lines. Electric power lines used to move large quantities of electricity at high voltages for long distances from the generator to the distribution lines of the local utility system.

Turbine wheel. A device, usually powered by steam or water, used to spin an electric generator.

Underground distribution. Electric distribution method using underground lines to enter the building. They are connected to transformers located either above or at ground level.

Volt. A unit of electric pressure. The pressure applied to force electrons through the circuit. The pressure that makes electrons move when an appliance starts or a light is turned on. Generally referred to as voltage, this pressure is available in live (hot) wiring circuits whether or not the equipment is turned on. Abbreviated "V".

Voltage. Force used to conduct electrons along a wire (conductor). As electrical current passes through a load, voltage drop results.

Voltmeter. A gauge or device for measuring volts (electrical pressure).

Water power. Renewable power supply from an generating plant that use water pressure from a reservoir to operate electric generating equipment.

Watt. A unit of measure for electric power. One watt of power equals one volt of pressure times one ampere of current. Abbreviated "w".

Watt hours. The number of watts of electricity used by a light, motor, heating or other electrical device in an hour. Abbreviated "wh", 1000 wh= 1 kilowatt hour (kwh).

Organizations Concerned with Electricity and Electrical Safety		
Acronym	Name of Organization	Responsibility
ANSI	American National Standards Institute	Electric safety standards
CSA	Canadian Standards Association	Electric safety standards
NESF	National Electric Safety Foundation	Electric safety standards
UL	Underwriters Laboratories	Electric safety standards
CPSC	Consumer Product Safety Commission	U.S. Govt.
EPA	Environmental Protection Association	U.S. Govt.
APPA	American Public Power Association	Electrical utilities
EEI	Edison Electric Institute	Electrical utilities
NRECA	National Rural Electrical Cooperative Association	Electrical utilities
AHAM	Association of Home Appliance Manufacturers	Trade association
IAEI	International Association of Electrical Inspectors	Trade association
IBEW	International Brotherhood of Electrical Workers	Trade association
NECA	National Electric Contractors Association	Trade association
NEMA	National Electric Manufacturers Association	Trade association
NFPA	National Fire Protection Association	Trade association
IES	Illuminating Engineering Society	Trade association
AACFS	American Association of Consumer and Family Sciences	Professional association
AAVIM	American Association for Vocational Instructional Materials	Educational association

Index

Credits

AAVIM wishes to express appreciation to the individuals and companies who participated as reviewers or contributed information and photos for the use of the publisher.

Without their help and involvement, publication of this book would have been impossible.

The information and materials supplied do not imply any endorsement by any individual, company or organization of the information presented within this publication

Consultants and Reviewers

Charles O. Shults, Retired Engineer and Department Manager; Warren RECC, Bowling Green, Kentucky

Ivan L. Winsett, Retired Engineer; Representative for Ronk Electrical Industries, Maxeys, Georgia

Reviewers

James M. Allison, Engineer and Professor; The University of Georgia, Athens, Georgia

Allen Anderson, Manager, Member Services Department; South Kentucky RECC, Somerset, Kentucky.

Roger H. Beckham, General Manager; Nashville Thermal Transfer Co., Nashville, Tennessee

Don Clodfelter, General Manager; Jackson County Rural Electric Membership Corporation, Brownstown, Indiana

Phillip H. Cox, Executive Director; International Association of Electrical Inspectors, Richardson, Texas

Ralph Duncan, Consultant, Instructor, and member of the National Code Committee; Columbus, Georgia

Alvin Etheredge, Engineer, Duke Power Company, Greensboro, North Carolina

Phillip M. Fravel, Professor; Virginia Polytechnic Institute and State University, Blacksburg, Virginia

Steven Metcalf, Indianapolis Power and Light, Indianapolis, Indiana

Ron Nelson, Manager; Member Services Department, Warren Rural Electric Cooperative Corporation, Bowling Green, Kentucky

John Orangenberg, Manager; Consumer Affairs Department, Underwriters Laboratories, Inc., Northbrook, Illinois

Mark A. Payne, Instructor; North Iredell High School, Olin, North Carolina

Charles Reed, State Specialist; Agriscience, Technical Education, Alabama State Department of Education, Montgomery, Alabama

James D. Kendrick, Coordinator; Curriculum, Resource and Evaluation Unit, Alabama State Department of Education, Montgomery, Alabama

Brian Wolka, Member Services Director; Jackson County Rural Electric Membership Corporation, Brownstown, Indiana

George F. Zelazny, Manager Product Safety Office; Sears Roebuck and Co., Hoffman Estates, Illinois

Contributors of Photographs and Information

American Wind Power Energy Association; Kent Robertson, Communications/Publications, Washington, DC

Athens Convention & Visitors Bureau; Peter Dale and Eric Holder, Athens, Georgia

Baldor Electric Company; G. Hubbard, Ft. Smith, Arkansas

Canadian Standards Association; Steve Mahoney, Communications Coordinator, Etobicoke, Ontario

Duke Power Company; Gary D. Barker, Community Relations Representative, Seneca, South Carolina

Ford Motor Company, Jim LaReese, Public Relations Department, Hapeville, Georgia

General Electric Lighting; R. J. Blazey, Manager; Graphic Communications, Cleveland, Ohio

David Gentry, Public Relations and Communications Representative; Warren RECC, Bowling Green, Kentucky

Georgia Power Company, Rick Ward and Curtis D. Hart, Atlanta, Georgia

Georgia Peanut Commission; Joan S. Underwood, Advertising/Production Manager, Tifton, Georgia

Georgia Milk Producers; William A. Moore, Executive Director, Atlanta, Georgia

Geothermal Education Office; Tiburon, California

Indianapolis Power and Light Co.; Steve Metcalf, Indianapolis, Indiana

Intertek Testing Services NA Inc.; Alan Dittrich, Andover, Massachusetts

Jackson County REMC, Betty Baute, Member Services Assistant, General Manager, Brownstown, Indiana

Nashville Thermal Transfer Company; Roger H. Beckham, General Manager, Nashville, Tennessee

National Electrical Safety Foundation; Donald A. Mader, Chairman, Rosslyn, Virginia

National Mining Association; Elizabeth Roca-Crooks, Editorial Assistant, Washington, DC
Nissan Motor Manufacturing Corporation USA; Kevin Gwin, Public Relations Specialist, Smyrna, Tennessee

Nissan Motor Manufacturing Corporation USA; Kevin R. Gwin, Public Relations Specialist, Smyrna, Tennessee

Nuclear Energy Institute; Publications Office, Washington, DC

PacifiCorp, Pacific Power/Utah Power; Margaret Kesler, Communications Department, Salt Lake City, Utah

St. Mary's Health Care System, Inc.; Lorraine Edwards, Public Relations Director, Athens, Georgia

The Watt Stopper, Inc.; Joy Cohen, Santa Clara, California

Underwriters Laboratories Inc.; Sandy Gentry, Media Relations Associate, Northbrook, Illinois

U.S. Army Corps of Engineers; John McKern, Walla Walla, Washington

United States Department of the Interior, Bureau of Reclamation, Kelly Conner, Visual Information Assistant, Bolder City, Nevada

Whirlpool Corporation; Suellen Russell, Communications Assistant, Benton Harbor, Michigan

Ivan L. Winsett, Maxeys, Georgia